Emerging Trends in Mechatronics

Series Editor

Aydin Azizi, Oxford, UK

Ajay Kumar · Parveen Kumar · Aydin Azizi
Editors

Trends in Mechatronics Systems

Industry 4.0 Perspectives

Springer

Editors
Ajay Kumar
Department of Mechanical Engineering
JECRC University
Jaipur, Rajasthan, India

Aydin Azizi
School of Engineering, Computing
and Mathematics
Oxford Brookes University
Oxford, UK

Parveen Kumar
Department of Mechanical Engineering
Rawal Institute of Engineering
and Technology
Faridabad, Haryana, India

ISSN 2731-4855 ISSN 2731-4863 (electronic)
Emerging Trends in Mechatronics
ISBN 978-981-97-9107-1 ISBN 978-981-97-9108-8 (eBook)
https://doi.org/10.1007/978-981-97-9108-8

This Springer imprint is published by the registered company Springer Nature Singapore Pte Ltd.
The registered company address is: 152 Beach Road, #21-01/04 Gateway East, Singapore 189721, Singapore

Preface

The term "mechatronics" came about as a made-up word from mechanics and electronics, where electronics means "hardware" and "software", and mechanics is the generic term for the disciplines of "mechanical engineering" and "hydraulics" so the aim is to achieve a synergistic optimization of mechanical engineering, electronic hardware and software in order to project more functions at lower cost, less weight and installation space, and better quality which makes the successful use of mechatronics in a problem solution. This objective can be achieved by integrating tools and strategies of Industry 4.0 with mechatronics systems. The integration of Industry 4.0 and mechatronics systems represents a powerful combination that brings together advanced technologies, data-driven insights, and intelligent automation. This book integrates the strategies of Industry 4.0 like artificial intelligence, Internet of Things, Digital Twins, etc., with mechatronics to maintain its sustainability. This book is an edited compilation of original research of mechatronics systems in context of Industry 4.0. This book has the following key features:

- Integrate Industry 4.0 tools with mechatronics systems.
- This book compiles overall aspects of advancements in mechatronics by bringing the latest research and development by comprehensive range of mathematical, numerical and simulation, modeling techniques of Industry 4.0 to strengthen the engineering science and technological developments for future.
- Discuss novel results of research and development and indicate the trends of development in the interdisciplinary field of Mechatronics systems.

The book consists of 12 chapters that discuss emerging trends in mechatronics systems in context of Industry 4.0.

- Chapter "Design and Fabrication of Automated Rivet Assembly and Pneumatic Punching Systems" develops a reliable and effective solution that improves production operations through the utilization of cutting-edge technologies such as CAD modeling, motion analysis and simulation software.

- Chapter "Real-World Security-Constrained AC Optimal Power Flow Analysis Using Artificial Gorilla Troops Optimizer" focuses on the utilization of an advanced algorithm, the Artificial Gorilla Troops Optimizer (AGTO), for scrutinizing some of the benchmark functions and addressing the real-world ACOPF problem.
- Chapter "Boosting Accuracy and Transparency in Biometric Finger-Vein Recognition: CNN with Adaptive Interpretable Recognition" contributes to the development of robust and transparent biometric recognition systems, with potential applications in security, access control and identity verification.
- Chapter "Design of Machine-Learning Framework for Cyber Attack Detection in Internet of Healthcare Things (IoHT): Critical Analysis" provides a machine learning framework for IoHT ecosystem cyber threat identification, including its design and implementation.
- Chapter "Optimizing Linear Regression model in Water Hardness Prediction for Industry 4.0" optimizes linear regression models for water hardness prediction using Industry 4.0 feature selection methods, particularly RFE.
- Chapter "Survey on Li-Fi: A Paradigm Shift in Data Transmission under VLC" discusses the overview of LiFi, its characteristics, a comparison of various data transfer technologies such as Wi-Fi, LiFi and its applications such as underwater data transmission, its usage in thermal power plants, hotels, hospitals, malls, etc.
- Chapter "Securing the Future: The Intersection of IoT, Cybersecurity Implementation, and AI Management" aims to identify and characterize the various categories of IoT applications, assess potential security risks related to them and their effects on these applications and present a linear programming methodology (LPP) for providing funding for multiple IoT and vulnerability projects.
- Chapter "Unlocking the Power of IoT, Machine Learning, and Cloud Computing: Case Studies in Smart Homes" contributes scientifically by identifying key technological implementations and their impacts and outlining future research directions to enhance smart home environments.
- Chapter "Comparative Analysis of Malware Detection Techniques and Machine-Learning Algorithms Used for Security Testing of Android Applications" provides means for understanding of malware detection techniques and analyze the machine learning algorithms used for security testing of Android applications.
- Chapter "A Multi-Biometrics System Based on the IRIS Using Convolutional Neural Networks" explains the presentation of trial assessment on iris pictures from the MMU iris data set.
- Chapter "Multi-objective Moth Flame Optimization for Renewable Energy Integration" utilizes a non-dominated multi-objective moth flame optimizer integrated with a fuzzy decision-making framework to balance supply and demand efficiently.
- Chapter "Improving Primary Storage Performance Using Data Deduplication and Pre-fetching Methods in Cloud" discusses how to reduce the client data download time and reduce the data access process.

The editors acknowledge the professional support received from Springer and express their gratitude for this opportunity.

Reader's observations, suggestions and queries are welcome.

Jaipur, India
Faridabad, India
Oxford, UK

Ajay Kumar
Parveen Kumar
Aydin Azizi

Acknowledgements

The editors are grateful to the Springer for showing their interest to publish this book in the buzz area of *Trends in Mechatronics Systems—Industry 4.0 Perspectives*. The editors express their personal adulation and gratitude to Loyola D'Silva (Publishing Editor) Springer, for giving consent to publish our work. He undoubtedly imparted the great and adept experience in terms of systematic and methodical staff who have helped the editors to compile and finalize the manuscript. The editors also extend their gratitude to Ms. Vinothini Elango (Project Coordinator—Total Service, Books Production), Springer, for supporting during her tenure.

The editors wish to thank all the chapter authors to contribute their valuable research and experience to compile this volume. The chapter authors, corresponding author in particular, deserve special acknowledgements for bearing with the editors, who persistently kept bothering them for deadlines, and with their remarks.

Dr. Ajay also wishes to express his gratitude to his parents, Sh. Jagdish and Smt. Kamla, and his loving brother Sh. Parveen for their true and endless support. They have made him able to walk tall before the world regardless of sacrificing their happiness and living in a small village. He cannot close these prefatory remarks without expressing his deep sense of gratitude and reverence to his life partner Mrs. Sarita Rathee for her understanding, care, support and encouragement to keep his moral high all the time. No magnitude of words can ever quantify the love and gratitude the he feels in thanking his daughters, Sejal Rathee and Mahi Rathee, and son Kushal Rathee who are the world's best children.

Finally, the editors obligate this work to the divine creator and express their indebtedness to the "almighty" for gifting them power to yield their ideas and concepts into substantial manifestation. The editors believe that this book would enlighten the readers about each feature and characteristics of *Trends in Mechatronics Systems—Industry 4.0 Perspectives*.

Ajay Kumar
Parveen Kumar
Aydin Azizi

Contents

Editors and Contributors

About the Editors

Ajay Kumar is currently serving as a Professor in Department of Mechanical Engineering, School of Engineering and Technology, JECRC University, Jaipur, Rajasthan, India. He received his Ph.D. in the field of Advanced Manufacturing from Guru Jambheshwar University of Science and Technology, Hisar, India after B.Tech. (Hons.) in mechanical engineering and M.Tech. (Distinction) in manufacturing and automation. His areas of research include Automation, Incremental Sheet Forming, Artificial Intelligence, Sustainable Materials, Additive Manufacturing, Mechatronics, Smart Manufacturing, Industry 4.0, Waste Management, and Optimization Techniques. He has over 150 publications in international journals of repute including SCOPUS, Web of Science and SCI indexed database and refereed international conferences. He has also co-authored and co-edited more than 20 books with ELSEVIER, CRC Press, WILEY, De Gruyter and IGI Global and conference proceedings with IOP. He has organized various national and international events including an International conference on Mechatronics and Artificial Intelligence (ICMAI-2021), International conference on Artificial Intelligence, Advanced Materials, and Mechatronics Systems (AIAMMS-2023) as conference chair. He has more than 20 national and international patents to his credit. He has supervised more than 8 M.Tech., Ph.D. scholars and numerous undergraduate projects/theses. He has a total of 13 years of experience in teaching and research. He is Guest Editors and Review Editor of the reputed journals including *Frontiers in Sustainability*. He has contributed to many international conferences/symposiums as a session chair, expert speaker and member of the editorial board. He has won several proficiency awards during the course of his career, including merit awards, best teacher awards and so on. He is adviser of QCFI, Delhi Chapter student cell at JECRC University and has also authored many in-house course notes, lab manuals, monographs and invited chapters in books. He has organized a series of Faculty Development Programs, International Conferences, workshops, and seminars for researchers, Ph.D., UG and

PG level students. He is associated with many research, academic and professional societies in various capacities.

https://scholar.google.co.in/citations?user=TmZS4JIAAAAJ&hl=en
https://www.webofscience.com/wos/author/record/D-5813-2019
Publons Profile: https://publons.com/researcher/1596469/ajay-kumar/
Research Gate: https://www.researchgate.net/profile/Ajay_Kumar349
ResearcherID: D-5813-2019
http://www.researcherid.com/rid/D-5813-2019

Parveen Kumar is currently serving as Assistant Professor and Head in Department of Mechanical Engineering at Rawal Institute of Engineering and Technology, Faridabad, Haryana, India. He has submitted Ph.D. from National Institute of Technology, Kurukshetra, Haryana, India. He completed his B.Tech. (Hons.) from Kurukshetra University, Kurukshetra, India and M.Tech. (Distinction) in Manufacturing and Automation from Maharshi Dayanand University, Rohtak, India. His areas of research include Intelligent Manufacturing Systems, Materials; Die less Forming, Design of Automotive systems, Additive Manufacturing, CAD/CAM and Artificial Intelligence, machine learning and Internet of Things in manufacturing, multi-objective optimization techniques. He has over 50 publications in international journals of repute including SCOPUS, Web of Science and SCI indexed database and refereed international conferences. He has 8 national and international patents in his credit. He has supervised two M.Tech. scholars and numerous undergraduate projects/theses. He has a total of 13 years of experience in teaching and research. He has co-authored/co-edited several books with ELSEVIER, CRC Press, WILEY, De Gruyter and IGI Global. He is also book series editor of *Industrial and Manufacturing Systems and Technologies: Sustainable and Intelligent Perspectives* in CRC Press, Taylor and Francis. He has organized a series of Faculty Development Programmes, workshops, and seminar for researcher, UG level students.

https://scholar.google.com/citations?hl=en&authuser=1&user=ESQ-RnY AAAAJ

Aydin Azizi holds a Ph.D. degree in Mechanical Engineering. Certified as an official instructor for the Siemens Mechatronic Certification Program (SMSCP), he currently serves as a Senior Lecturer at the Oxford Brookes University. His current research focuses on investigating and developing novel techniques to model, control and optimize complex systems. Dr. Azizi's areas of expertise include Control and Automation, Artificial Intelligence and Simulation Techniques. Dr. Azizi is the recipient of the National Research Award of Oman for his AI-focused research, DELL EMC's "Envision the Future" completion award in IoT for "Automated Irrigation System" and "Exceptional Talent" recognition by the British Royal Academy of Engineering.

https://scholar.google.com/citations?user=CoQco90AAAAJ&hl=en

Contributors

Laith Abualigah Computer Science Department, Al al-Bayt University, Mafraq, Jordan;
MEU Research Unit, Middle East University, Amman, Jordan;
Applied Science Research Center, Applied Science Private University, Amman, Jordan;
Centre for Research Impact and Outcome, Chitkara University Institute of Engineering and Technology, Chitkara University, Rajpura, Punjab, India;
Artificial Intelligence and Sensing Technologies (AIST) Research Center, University of Tabuk, Tabuk, Saudi Arabia

Sourabh Anand Communication and Technology, University School of Information, Guru Gobind Singh Indraprastha University, New Delhi, India

Arpna Department of Chemistry, Baba Mastnath University Asthal Bohar, Rohtak, Haryana, India

Pushpendra S. Bharti Communication and Technology, University School of Information, Guru Gobind Singh Indraprastha University, New Delhi, India

M. Bhavya Department of AI & ML, M S Ramaiah Institute of Technology, Bangalore, India

Buddesab Department of Engineering, APJ Abdul Kalam School of Engineering, Garden City University, Banaglore, India

Shankar Chakraborty Production Engineering Department, Jadavpur University, Kolkata, India

Nupa Ram Chouhan College of Computing Sciences and Information Technology, Teerthanker Mahaveer University, Moradabad, Uttar Pradesh, India

Neeraj Dahiya Department of Computer Science and Engineering, SRM University, Sonipat, Haryana, India

Surjeet Dalal Department of CSE, Amity University, Gurugram, Haryana, India

G. Geetha Department of Artificial Intelligence and Machine Learning, Cambridge Institute of Technology, Bengaluru, KAR, India

Archna Goyal K R Mangalam University, Gurugram, Haryana, India

Mohammad Haider College of Computing and Informatics, Saudi Electronic University, Riyadh, Saudi Arabia

Syed Hayath Department of Artificial Intelligence and Machine Learning, Cambridge Institute of Technology, Bengaluru, KAR, India

Pradeep Jangir Department of Biosciences, Saveetha School of Engineering, Saveetha Institute of Medical and Technical Sciences, Chennai, India; Jadara Research Center, Jadara University, Irbid, Jordan

Kanak Kalita Department of Mechanical Engineering, Vel Tech Rangarajan Dr. Sagunthala R&D Institute of Science and Technology, Chennai, India; University Centre for Research and Development, Chandigarh University, Mohali, India

Manvi Khatri Department of Computer Science and Engineering, SRM University, Sonepat, India

Ajay Kumar Department of Mechanical Engineering, School of Engineering and Technology, JECRC University, Jaipur, Rajasthan, India

Mukul Kumar Communication and Technology, University School of Information, Guru Gobind Singh Indraprastha University, New Delhi, India

Parveen Kumar Department of Mechanical Engineering, Rawal Institute of Engineering and Technology, Faridabad, Haryana, India

Pawan Kumar Department of Computer Science and Engineering, Deenbandu Chhotu Ram University of Science and Technology, Murthal, India

Santosh Kumar Department of Computer Science and Engineering, Galgotias University, Greater Noida, Uttar Pradesh, India

Vikas Kumar Department of Computer Science and Engineering, World College of Technology and Management, Gurgaon, Haryana, India

Kamlesh Lakhwani JECRC University, Jaipur, Rajasthan, India

Nikhil Department of Chemistry, Baba Mastnath University Asthal Bohar, Rohtak, Haryana, India

Sundaram B. Pandya Department of Electrical Engineering, Shri K.J. Polytechnic, Bharuch, India

Ashish Payal Communication and Technology, University School of Information, Guru Gobind Singh Indraprastha University, New Delhi, India

K. C. Rajani Department of Engineering, APJ Abdul Kalam School of Engineering, Garden City University, Banaglore, India

Sangeeta Rani Department of Computer Science and Engineering, World College of Technology and Management, Gurgaon, Haryana, India

Tanvi Rustagi Department of Computer Science and Engineering, World College of Technology and Management, Gurgaon, Haryana, India

Anuj Shakya ABES Engineering College, Ghaziabad, India

Ajay Sharma Department of Computer Science and Engineering, SRM University, Sonepat, India

Bhavna Sharma JECRC University, Jaipur, Rajasthan, India

Satyajeet Sharma JECRC University, Jaipur, Rajasthan, India

Sidhu Department of Computer Science and Engineering, Galgotias University, Greater Noida, Uttar Pradesh, India

Sukhdip Singh Department of Computer Science and Engineering, Deenbandu Chhotu Ram University of Science and Technology, Murthal, India

Surbhi College of Computing Sciences and Information Technology, Teerthanker Mahaveer University, Moradabad, Uttar Pradesh, India

J. Thriveni University of Visvesvaraya College of Engineering, Bangalore University, Bangaluru, KAR, India

K. R. Venugopal Bangalore University, Bangaluru, KAR, India

Chanchal Yadav ABES Engineering College, Ghaziabad, India

Design and Fabrication of Automated Rivet Assembly and Pneumatic Punching Systems

Mukul Kumar, Sourabh Anand, Parveen Kumar, Pushpendra S. Bharti, Ashish Payal, and Ajay Kumar

Abstract The process of manually riveting during the manufacturing of an electromagnetic relay frequently results in irregularities and faults, which impede both efficiency and the quality of the final product. Automation is crucial in addressing these issues, particularly in the areas of rivet assembly and pneumatic punching. This research aims to automate this crucial process with a specific emphasis on enhancing productivity. Initially, a Computer-Aided Design (CAD) model is generated using Fusion 360 software. Subsequently, a physical prototype is fabricated. Precise selection of essential components is carried out, and thorough construction of mechanical, electrical, and pneumatic parts is undertaken, adhering to the specifications provided by the CAD models. Throughout the creation process, the system accurately determines specific requirements and combines different components, such as motors, pneumatic punches, frames, and sensors, into a cohesive and operational entity. CAD simulations and programming tools are employed to enhance the efficiency of designs. A prototype is created as a foundation for implementing automated riveting assembly in industry.

Keywords Manufacturing · Automation · Productivity · Computer-Aided Design · Sensors · Simulations

M. Kumar · S. Anand · P. S. Bharti · A. Payal
Communication and Technology, University School of Information, Guru Gobind Singh Indraprastha University, Sector 16C, Dwarka, New Delhi 110078, India

P. Kumar
Department of Mechanical Engineering, Rawal Institute of Engineering and Technology, Faridabad, Haryana, India

A. Kumar (✉)
Department of Mechanical Engineering, School of Engineering and Technology, JECRC University, Jaipur, Rajasthan, India
e-mail: ajay.kumar30886@gmail.com

A. Kumar et al. (eds.), *Trends in Mechatronics Systems*, Emerging Trends in Mechatronics, https://doi.org/10.1007/978-981-97-9108-8_1

1 Introduction

In the ever-changing world of contemporary manufacturing, where the constant improvement of production processes is crucial for achieving efficiency, accuracy, and cost-effectiveness, the assembly of components plays a crucial role. In the manufacturing process of electromechanical relays, the crucial step involves manually installing rivets followed by pneumatic punching. This approach is widely employed in several industries such as automotive, aerospace, and electronics. Workers traditionally align each rivet with great precision and use pneumatic punching machines to securely attach them throughout the process of hand assembly which is shown in the Fig. 1. Nevertheless, this approach, while efficient, has inherent disadvantages such as the potential for human mistake, discrepancies in the quality of rivet placement, and longer cycle durations. Furthermore, the dependence on human labour presents difficulties in expanding production to meet growing requirements, emphasizing the urgent want for automated alternatives. This research programme seeks to address these issues by creating and deploying an automated system for rivet assembly and pneumatic punching. By utilizing developments in robotics and pneumatic technology, the system aims to transform the assembly process by improving the accuracy of rivet placement and speeding up punching processes. By using accurate sensors, the system guarantees to enhance production efficiency and minimize errors, thereby revolutionizing the manufacturing industry. Implementing automation in the rivet assembly and punching process offers significant advantages for firms, including decreased lead times, improved product uniformity, and heightened manufacturing adaptability. Moreover, automation enables the reallocation of human resources to activities that need cognitive abilities, thus cultivating a workforce proficient in innovation and analytical thinking. The following sections of this study explore the complexities of the proposed automated system, providing a full description of its methodology, aims, and expected workflow. An extensive examination of relevant literature offers valuable knowledge about the latest research and technological progress, placing the proposed project in the wider framework of factory automation. This programme seeks to advance factory automation by carefully creating and implementing solutions with the goal of producing substantial improvements in production processes.

2 Importance of Automation

The significance of automation in contemporary production, as outlined in Table 1, encompasses various crucial facets. First and foremost, automation profoundly alters industrial processes by completely changing how goods are conceived, produced, and delivered. This change results in substantial enhancements in efficiency, precision, and scalability throughout production lines. Additionally, automation decreases dependence on manual labour, thereby eliminating human errors and guaranteeing

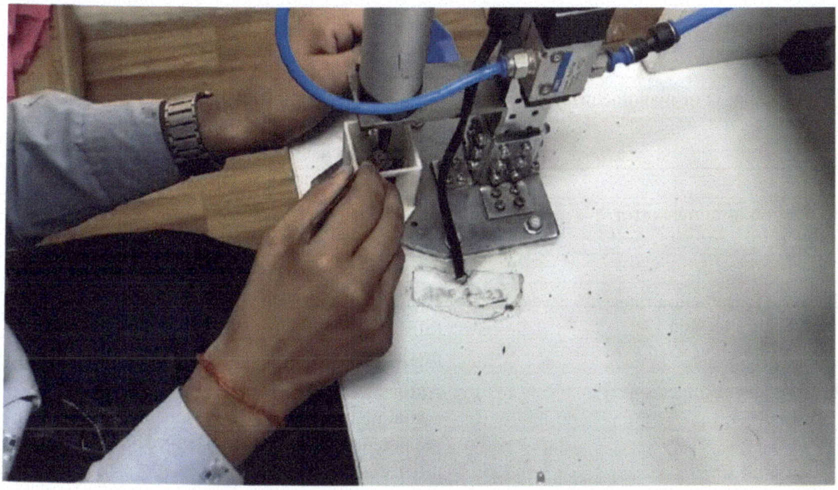

Fig. 1 Manually riveting by worker using pneumatic punch

constant manufacturing quality. Furthermore, automated methods facilitate higher production rates, allowing for the fulfilment of large-scale requirements while maintaining product quality. Morcovcr, automation improvcs workplacc safcty by assuming control of perilous jobs, thus diminishing the likelihood of casualties linked to manual procedures. Furthermore, the incorporation of automation results in sustained financial benefits by increasing efficiency and decreasing operational costs. Moreover, automation enables manufacturers to monitor and gather data in real-time, allowing them to improve processes, predict maintenance requirements, and quickly adjust to fluctuations in market demand [1–5]. Furthermore, the continuous technological progress that is propelling the importance of automation in manufacturing highlights its function in promoting innovation and sustaining competitiveness in the industry [6, 7]. These several elements jointly emphasize the profound influence of automation on contemporary industrial methods.

3 Literature Review

In 2004, Messler conducted a study on the historical development and current progress of joining technologies. The study provided a thorough examination of the many methods used to join materials, which laid the foundation for further research on specialized applications like riveting [12]. Dasari (2012) examined the mechanical characteristics of constructions including riveted connections, specifically studying the impact of rivets on structural damping. This study emphasized the practical consequences of improving the strength and stability of different engineering applications [13]. Xi et al. (2013) presented an innovative framework for

Table 1 Importance of automation in modern manufacturing

Importance	Description
Transformation of manufacturing processes	Automation fundamentally alters the process of creation, manufacture, and distribution, completely changing production lines by improving efficiency, accuracy, and scalability [8]
Reduction of human error	Automation decreases dependence on manual labour, decreasing human errors and guaranteeing constant manufacturing quality [8]
Increased production rates	Facilitates accelerated manufacturing rates and fulfils large-scale requirements while maintaining product quality standards [9]
Enhanced workplace safety	Automated systems enhance workplace safety by assuming hazardous activities, hence diminishing the potential for injuries linked to manual processes [10]
Cost savings	The integration of automation results in long-term cost savings by increasing productivity and reducing operational expenses [5]
Future significance and innovation	The increasing significance of automation in manufacturing, driven by technological advancements, will foster innovation and sustain competitiveness within the industry [11]

robotic percussive riveting, which represents a notable progress in the field of factory automation. Their research presented a specialized robotic system designed specifically for the purpose of airplane assembly, with the objective of optimizing manufacturing processes and enhancing overall efficiency [14]. Sontakke et al. (2014) conducted a detailed analysis of drilling cum riveting machines, with a focus on improving the design and performance of hybrid machines that can perform both drilling and riveting tasks. These machines are important for achieving high precision in manufacturing for the aerospace and automotive industries [15]. Xi et al. (2017) further developed automated riveting technology by creating a more sophisticated robotic system specifically designed to improve the precision and dependability of riveting procedures in the production of airplanes. This study emphasized the continuous endeavours to include robotics into assembly lines to enhance productivity and provide quality control [16]. In 2019, Pina Cipriano focused on enhancing energy efficiency by optimizing the utilization of energy in friction-riveting procedures. The objective was to decrease manufacturing expenses and minimize the environmental footprint, all while ensuring the maintenance or enhancement of joint quality [17]. Bagherpour et al. (2019) utilized the analytic hierarchy approach to assess the effectiveness and dependability of self-piercing riveted joints in different industrial contexts. Their study offered a systematic framework for evaluating the performance and reliability of self-piercing riveting procedures [18]. Zdravkovic and Golovashchenko (2020) performed an extensive examination of self-piercing riveting, with a specific emphasis on the strength of the joint and the effectiveness of joining sheet metals. Their research provided useful insights into maximizing the utilization of self-piercing riveting to improve structural integrity and production efficiency [19]. Ribas Matas et al. (2021) recently conducted a study where they examined several rivet arrangement patterns to assess their efficacy in mending and strengthening aircraft structures. The purpose of this comparison investigation was to determine the most effective rivet configurations for enhancing structural strength and durability in aeronautical maintenance procedures [20]. Das et al. (2023) examined the incorporation of embedded anchoring in multi-material assemblies through friction-riveting processes. They investigated novel methods to effectively link different materials and improve assembly efficiency and structural performance in advanced manufacturing applications [21]. These studies which we discussed in Table 2 highlight the continuous development of rivet assembly technologies, starting from basic research on joining methods to advanced applications in automated systems, optimization of energy efficiency, and strategies for assembling multiple materials. Every contribution in this subject tackles important difficulties and makes progress, which will lead to future innovations in manufacturing and engineering processes.

4 Details About Electromechanical Relay

4.1 Production of Electromechanical Relay

The process of manufacturing an electromechanical relay which is shown in Fig. 3a commences with the essential tasks of riveting and punching, which are carried out manually by highly skilled technicians. Riveting is used to firmly fasten the metal pieces of the frame or enclosure, guaranteeing its structural strength. On the other hand, punching is employed to make exact holes and slots, which are necessary for accurately aligning and mounting following components. The involvement of human knowledge guarantees utmost care during these preliminary phases. Subsequently, the electromagnetic coil and core assembly are utilized, carefully produced by operators. The coil is coiled by hand around a core made of a ferromagnetic material, such as iron or steel, in order to achieve specific electrical properties, such as resistance and magnetic flux density. The keen attention to detail and precise manual modifications made by the personnel guarantee the coil's best functioning. The subsequent phase entails the production of the contacts, which include both fixed and mobile components. These contacts are made from materials that are both extremely conductive and resilient. Skilled technicians physically shape and finish them to reduce resistance and guarantee dependable performance after numerous switching cycles. Proficient artistry is essential to uphold the elevated criteria demanded. Once the coil, core, and contacts are prepared, the process of assembling the mechanical components commences. Workers manually build the frame and enclosure components. Subsequently, the coil and core assembly are thoroughly inserted into the frame.

Table 2 Literature of rivet assembly

Author(s) and Year	Title	Key focus	Key findings
Das et al. [21]	Embedded anchoring of multi-material assemblies by friction-riveting process	Friction-riveting applications	Investigates the use of embedded anchoring in multi-material assemblies
Ribas Matas et al. [20]	Comparison of rivet layout patterns in aircraft repair patches	Rivet layout comparison	Evaluates the efficacy of various rivet layouts
Zdravkovic and Golovashchenko [19]	Analysis of sheet metal joining with self-piercing riveting	Joint strength and efficiency	Investigation of the application of self-piercing riveting in the merging of sheet metal
Bagherpour et al. [18]	Assessment of self-piercing riveted joints using analytic hierarchy process	Performance evaluation	Assesses riveted connections that are self-piercing
Pina Cipriano [17]	Energy efficiency based analysis and optimization of friction riveting	Friction-riveting optimization	Enhances the efficacy of energy in friction riveting
Xi et al. [16]	A robotic percussive riveting system for aircraft assembly automation	Automation technology	Presents a robotic system for aircraft assembly
Sontakke et al. [15]	Design and analysis of drilling cum riveting machine	Machine design and performance	Detailed analysis of drilling riveting machines
Xi et al. [14]	Framework on robotic percussive riveting for aircraft assembly automation	Robotic percussive riveting	Robotic riveting framework for aircraft assembly
Dasari [13]	Damping of structures with riveted joints	Structural damping	Investigates the effect of damping on buildings utilizing rivets
Messler [12]	Joining of materials and structures: from pragmatic process to enabling technology	Evolution of joining technologies	Overview of the technologies used to join materials

The contacts and armature, along with their corresponding springs and mechanical linkages, are manually attached. The assembly is attentively executed to guarantee accurate alignment and seamless mechanical functioning, relying on the proficiency of the assembly professionals. Following insulation and sealing, insulating materials are physically applied to prevent electrical leakage, hence ensuring safety and reliability. The workers apply a seal to the relay in order to safeguard the internal components from pollutants such as dust and moisture, which might potentially undermine its operation. Human interaction makes certain that each component is properly insulated and tightly sealed. Prior to reaching the final stages, every relay goes through extensive testing and quality control procedures, conducted by specialists in quality assurance. This entails the manual verification of electrical and mechanical performance, as well as testing for factors like as contact resistance, insulation resistance, operating speed, and endurance under specific conditions. After undergoing testing, the relays are packaged and labelled manually, in accordance with their unique requirements and intended use. The necessary information, including part numbers, ratings, and production dates, is provided for the purpose of identification and traceability. Workers take care to ensure that proper labelling and secure packing are maintained. The last inspection phase entails a detailed manual examination conducted by quality inspectors to detect any flaws or deviations from specifications, guaranteeing that all relays adhere to the most stringent quality criteria prior to shipment or integration into larger systems. Ultimately, the relays can be customized to fulfil specific requirements for automotive, industrial control, or telecommunications applications. Proficient technicians make manual adjustments to characteristics such as voltage ratings, contact designs, and environmental durability in order to fulfil customer specifications. The Flowchart of Production of Electromechanical Relay is illustrated in Fig. 2.

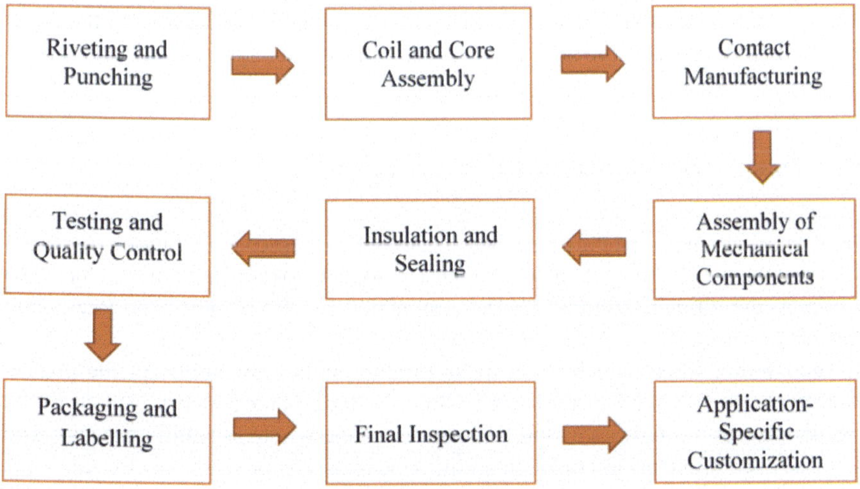

Fig. 2 Flowchart of production of electromechanical relay

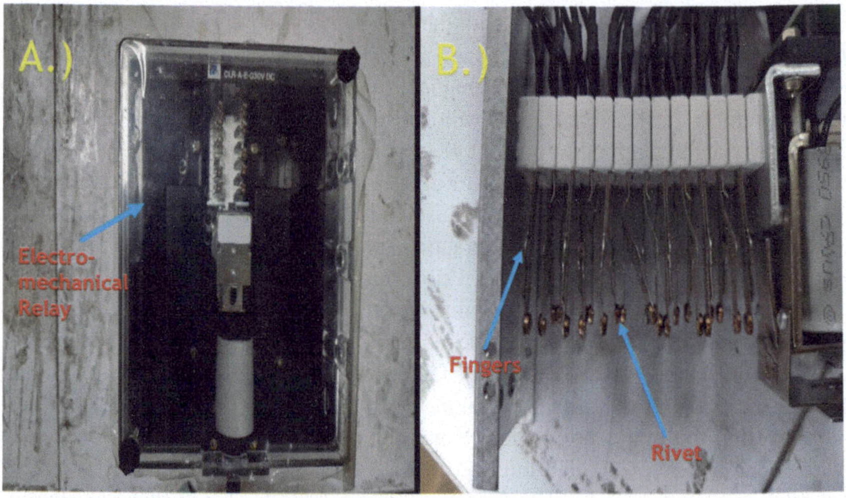

Fig. 3 **a** Electromechanical relay and **b** interior view of electromechanical relay

During the initial stage of riveting and punching, our primary objective is to accurately join a finger and rivet by utilizing a pneumatic punch. Within the interior view of the electromagnetic relay, there are clearly visible components known as fingers and rivets, as depicted in Fig. 3b. The assembly procedure is vital because it produces the finger and rivet component, which is essential for the contact configuration of the electromechanical relay. More precisely, the contact arrangement is in the normally open (NO) state when the rivet's completed part is not in touch with the other part of the relay. On the other hand, the arrangement is typically classified as closed (NC) if the rivet's assembled part is in contact with the other part. It is crucial to guarantee the precision of this assembly in order to ensure the dependable functioning of the relay, since it directly impacts the relay's switching behaviour and overall performance.

4.2 Design of Automated Rivet Assembly

Our design has simplified the assembly procedure by incorporating only two essential components: the riveting wheel and the base plate. The streamlined design has been chosen as the ultimate solution for the Automated Rivet Assembly and Pneumatic Punching project.

The riveting wheel which is shown in the Fig. 4a is responsible for the precise placement and fastening of the rivets, while the base plate which is shown in the Fig. 4b ensures a robust and solid foundation for the whole assembly process. The decrease in components not only streamlines the assembly process but also improves the efficiency and dependability of the system, facilitating its manufacturing, maintenance, and operation. This ultimate design which is shown in Fig. 4c efficiently tackles the difficulties faced in prior models and delivers an optimized solution for improving efficiency in production operations.

In the riveting wheel, grooves are precisely created to guarantee that the finger is securely placed in one groove, while another groove is earmarked for the rivet as

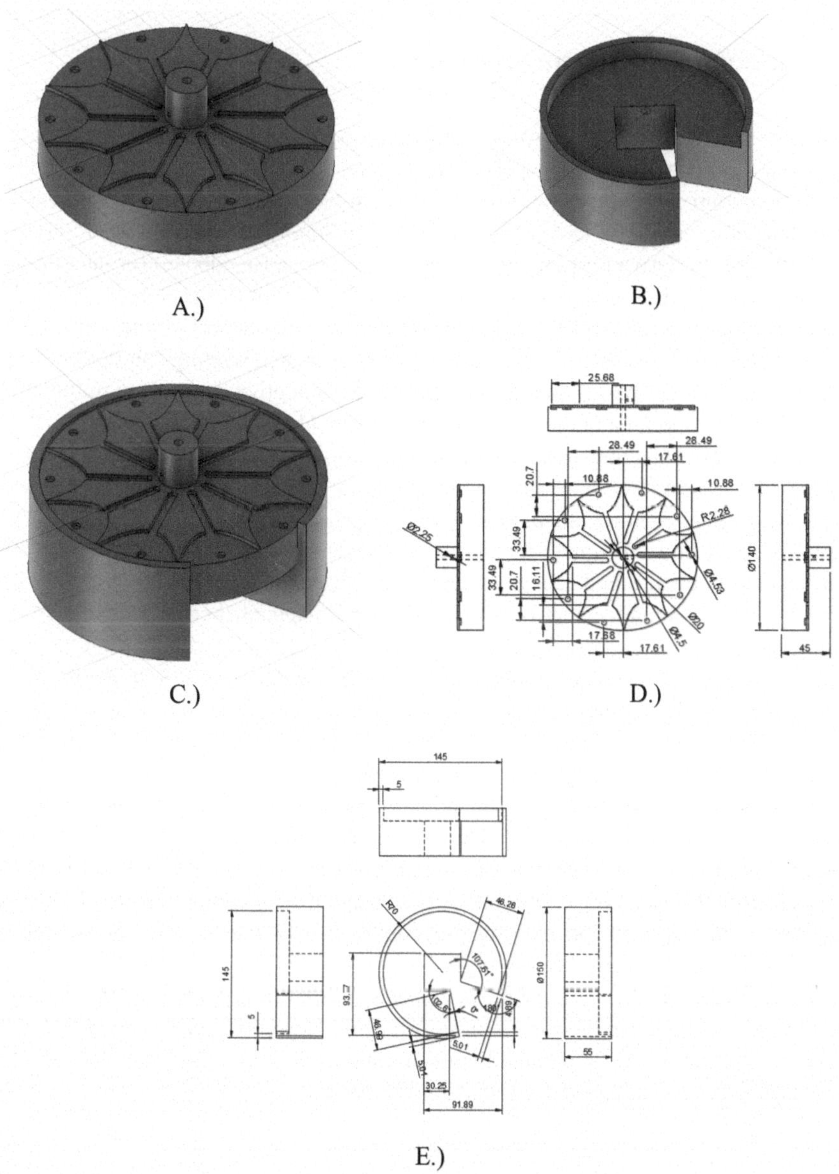

Fig. 4 **a** Riveting wheel, **b** base plate, **c** assembly of prototype, **d** 2D drawing of riveting wheel and **e** 2D drawing of base plate

shown in the Fig. 4d. Also, the rivet is placed in the groove that is designated for it. Because of this perfect design, the positioning of both components can be done with greater ease, which guarantees that the assembly process will be both dependable and effective. 2D Drawing of Base Plate as shown in the Fig. 4e.

5 Automated—Riveting Assembly

We have created a prototype Automated Riveting Assembly that includes various important components: a Stepper Motor for precise control of movement, an Arduino Uno microcontroller for programming logic, an Inductive Proximity Sensor for accurate detection, an SMPS (Switched Mode Power Supply) for stable power supply, an A4988 Motor Drive for powering the stepper motor, a Breadboard for integrating circuits, and an LED indicator for visual feedback. Within this prototype, the LED functions as a replacement for a pneumatic punch, as dictated by practical limitations. This showcases our inventive strategy for automated riveting solutions during the developmental phase. Integrating a proximity inductive sensor into automated rivet assembly and pneumatic punching systems yields substantial advantages. The sensor enables contactless detection, ensuring minimal wear and maintaining cleanliness while accurately detecting metal fingers in grooves. This enhances product reliability by preventing misfires and ensuring precise assembly. With its swift response time and robustness against dust and temperature variations, the sensor further boosts operational efficiency and safety. It enhances production quality by activating the pneumatic punch only when a finger is detected, thereby reducing errors and ensuring consistent, high-performance output. Meanwhile, stepper motors offer precise control and repeatable movements, achieving accurate positioning crucial for uniform assembly quality. Paired with feedback from sensors like inductive proximity sensors, stepper motors automate systems effectively, enhancing efficiency and reliability while minimizing human intervention.

Upon detecting a metal finger, the inductive proximity sensor activates the LED, indicating the stepper motor to stop. During this brief interval, lasting for a duration of 1 s, the illuminated LED (which is shown in Fig. 5a and b) It signifies the ideal timing for executing the punching operation. This configuration guarantees accurate timing and synchronization in the riveting procedure, in accordance with the operating specifications of the assembly prototype.

The final assembly, seen in Fig. 6, consists of a riveting wheel, a sensor frame that houses an inductive proximity sensor for finger detection, an assembly frame to hold the riveting wheel, a pneumatic punch with a solenoid valve, and an assembly base utilized as a foundation for the assembly process.

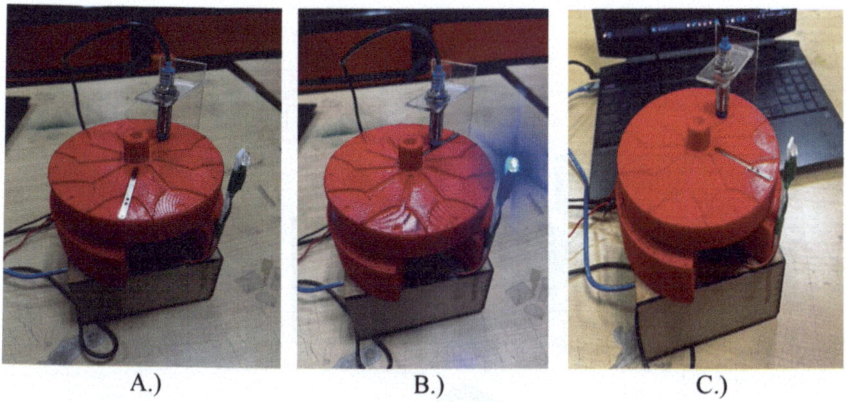

Fig. 5 **a** Before sensing the metal finger, **b** After sensing the metal finger and **c** Full assembly of the prototype.

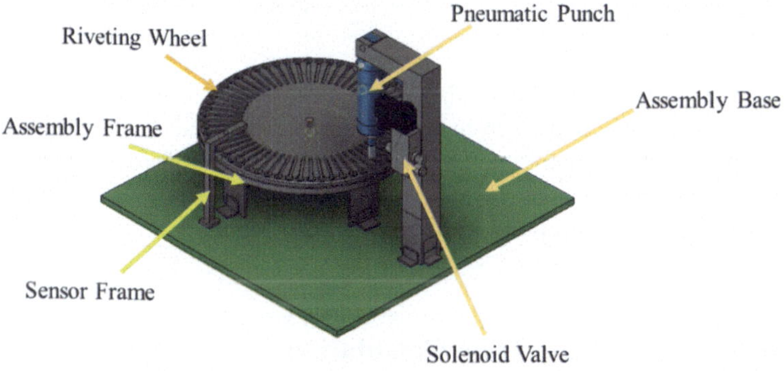

Fig. 6 Final design of assembly

5.1 Circuit Diagram

An Arduino Uno microcontroller, a 12 V switched mode power supply (SMPS), a NEMA 17 stepper motor, an inductive proximity sensor, a 5 V breadboard power supply, a light bulb, and an A4988 motor driver are all included in the circuit design for the project. Additionally, the project includes a light bulb. As a reliable supply of power for the circuit, the 12 V SMPS is serving its purpose. Managing the operation of the entire system is the responsibility of the Arduino Uno, which serves as the central control unit. The NEMA 17 stepper motor is responsible for providing accurate mechanical control, while the inductive proximity sensor is responsible for determining whether or not metallic items are present. The breadboard power supply, which operates at 5 V, guarantees that low-voltage components receive a suitable amount of power. In the circuit, the light bulb acts as a load that is indicative

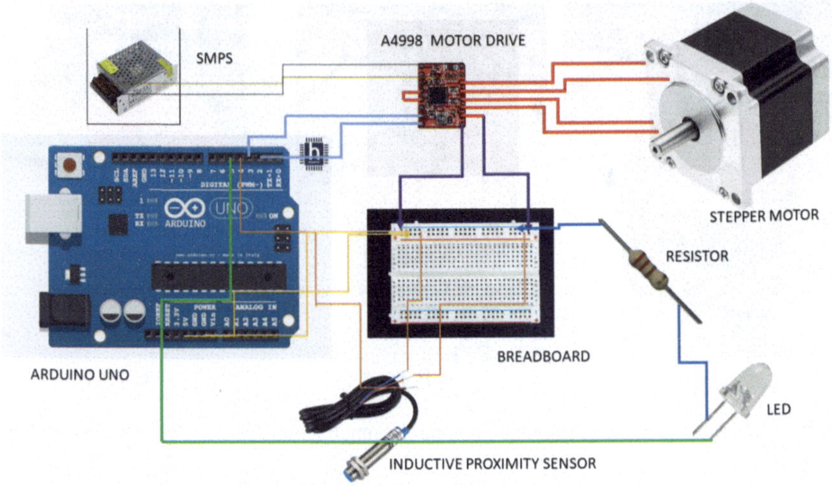

Fig. 7 Flowchart of implementing collaborative robots with RoboDK workstations

of the circuit. The final component, the A4988 motor driver, is responsible for facilitating the translation of signals from the Arduino into accurate motor motions. This ensures that the stepper motor operates without any hiccups. These components are connected to one another in accordance with the circuit diagram, which enables the project to perform as it was purposefully designed to (Fig. 7).

6 Bottle Filling Cycle Time Calculation

Within our automated riveting assembly prototype, the sensor input and LED control function in the following manner: upon detecting the presence of metal, the inductive proximity sensor activates the LED, causing it to illuminate for a duration of 1 s (delay (1000)). In terms of motor control, the stepper motor is instructed to complete a complete revolution (consisting of 15 steps) if metal is detected. Each step is accompanied by a delay of 11 ms (equivalent to 11,000 μs).

6.1 Calculating the Time for One Riveting Cycle

Time for LED Operation = 1 s.

For finding out the Time for Motor Operation, the stepper motor requires 2 * stepDelay microseconds for each step, which includes one high pulse and one low pulse. The motor performs a complete revolution in a total of 15 steps.

Time for one full revolution:

- Time per step = 2 × 11,000 μs= 22,000 μs= 0.022 s
- Total time for 15 steps = 15 × 0.022 s = 0.33 s

Total Cycle Time

- Total cycle time for one riveting operation:
- Total cycle time = Time for LED operation + Time for motor operation
- Total cycle time = 1 s + 0.33 s = 1.33 s

6.2 Comparing Manual Versus Automated Efficiency

Based on the worker's industry experience, the manual process requires 5 s every cycle.

Manual throughput: $N_{\{manual\}} = \frac{1}{T_{\{manual\}}} = \frac{1}{5} = 0.2$ cycles/s.

Automated throughput: $N_{\{auto\}} = \frac{1}{T_{\{auto\}}} = \frac{1}{1.33} = 0.75$ cycles/s.

Energy Consumption (E): In general, automation is usually more energy-efficient per cycle because it performs processes faster. However, to ensure accuracy, particular power consumption data is required. Automation generally enhances precision and decreases maintenance expenses by consistently and reliably performing operations, hence decreasing errors caused by human variability and ensuring the durability of equipment with reduced maintenance needs.

- ***Efficiency Calculation***

In order to ensure a fair comparison, we standardize the measures. Using hypothetical values to establish a baseline for normalization (with manual as the base and automated as the improvement):

- Throughput: $T_{\{manual,norm\}} = \frac{1.33}{5} = 0.266$, $T_{\{auto,norm\}} = 1$
- Accuracy: $A_{\{manual,norm\}} = 0.9$, $A_{\{auto,\ norm\}} = 1$
- Energy: $E_{\{manual\ ;\ norm\}} = \frac{50}{30} = 0.6$, $E_{\{auto;norm\}} = 1$
- Maintenance: $M_{\{manual\ ,norm\}} = 0.5$, $M_{\{auto,norm\}} = 1$

- ***Weighted Efficiency Score***

Determine the relative significance of each component by assigning weights.

- Weights: $w_T = 0.3$, $w_N = 0.3$, $w_E = 0.2$, $w_A = 0.1$, $w_M = 0.1$

$$\text{Efficiency}_{\{manual\}} = w_T \cdot N_{\{manual\ norm\}} + w_N \cdot N_{\{manual\ norm\}} + w_E \cdot E_{\{manual\ norm\}} + w_A \cdot A_{\{manual\ norm\}} + w_M \cdot M_{\{manual\ norm\}}$$

$$\text{Efficiency}_{\{auto\}} = w_T \cdot T_{\{auto\ norm\}} + w_N \cdot N_{\{auto\ norm\}} + w_E \cdot E_{\{auto\ norm\}}$$

$$+ w_A \cdot A_{\{\text{auto norm}\}} + w_M \cdot M_{\{\text{auto norm}\}}$$

$$\begin{aligned}\text{Efficiency}_{\{\text{manual}\}} &= 0.3 \times 0.266 + 0.3 \times 0.9 + 0.2 \times 0.6 + 0.1 \times 0.9 + 0.1 \times 0.5 \\ &= 0.6078\end{aligned}$$

$$\text{Efficiency}_{\{\text{auto}\}} = 0.3 \times 1 + 0.3 \times 1 + 0.2 \times 1 + 0.1 \times 1 + 0.1 \times 1 = 1$$

- ***Improvement***

$$\Delta\text{Efficiency} = \text{Efficiency}_{\{\text{auto}\}} - \text{Efficiency}_{\{\text{manual}\}} = 1 - 0.6078 = 0.3922$$

7 Discussions and Future Directions

7.1 Discussions Regarding the Implementation of Riveting Assembly

The study is a notable breakthrough in the field of industrial automation. The objective is to enhance manufacturing processes by incorporating automated methods for rivet assembly and pneumatic punching. The initiative seeks to address inherent challenges faced in traditional assembly processes, such as the requirement for manual labour, inefficiencies in manufacturing, and the potential for human errors. The research talks encompass a wide range of topics, starting with the idea and design of automated assembly systems. This involves careful assessment of factors such as component compatibility, system integration, and operational requirements. Furthermore, the study investigates the process of selecting appropriate components that are essential for constructing the automated systems. This includes pneumatic actuators, sensors, controllers, safety devices, and specialized tools required for performing riveting and punching jobs. When selecting materials, it is essential to consider factors such as their ability to endure stresses, their durability, and their compatibility with production processes. The inclusion of control software is another important aspect of the project discussions. Thorough testing and fine-tuning are necessary to guarantee smooth operation and reliability in real manufacturing environments. The study focuses on implementing testing methodologies to validate the performance, reliability, and safety of the automated assembly and punching systems. Deficiencies or constraints identified during testing are addressed through iterative refinement procedures, which may involve making changes to the design, manufacturing methods, or software algorithms. Furthermore, the debates also encompass the enhancement of the assembly process to attain utmost efficiency and productivity. This entails the identification of potential challenges, such as the complexity of components, and the enhancement of workflow methods to minimize downtime and enhance throughput.

Furthermore, rigorous consideration should be devoted to the development of training programmes for operators and maintenance personnel to ensure the proper operation, maintenance, and troubleshooting of the automated systems. Essentially, the project discussions revolve around a comprehensive plan to leverage automation technology in order to enhance efficiency in manufacturing operations. The project aims to transform conventional assembly methods, accelerate production processes, and fulfil industry demands for enhanced productivity, quality, and safety standards through the utilization of automated technology for rivet assembly and pneumatic punching.

7.2 Future Trends and Advancements in Prototype

The study has considerable promise for future advancements and contributions to industrial automation. Thanks to continuous technological advancements, several possibilities for expansion and enhancement exist. An effective strategy involves integrating advanced sensor technologies and machine learning algorithms to enhance the system's real-time monitoring capabilities, fault detection, and predictive maintenance capabilities. Furthermore, exploring the integration of collaborative robotics, also known as cobots, could provide a safer and more flexible connection between human workers and the automated assembly system, hence improving efficiency and productivity. Furthermore, there is a potential for incorporating additive manufacturing techniques to customize and enhance components for specific purposes, hence enhancing the overall efficiency and flexibility of the system. Given the continuous advancement of industries towards smart manufacturing and Industry 4.0 paradigms, this initiative has the capacity to greatly influence the future of manufacturing processes. This can be accomplished by employing sophisticated technologies and innovative automation techniques.

8 Conclusions

The study is a significant advancement in industrial automation. Through planning, extensive experimentation, and ongoing refinement, we have developed a reliable and effective solution that improves production operations. Through the utilization of cutting-edge technologies such as CAD modelling, motion analysis, and simulation software, we have effectively surmounted challenges and achieved an exceptional degree of precision in both the placement and performance of components. As a result, our automated method improves efficiency, reduces the reliance on manual labour, decreases the likelihood of errors, and ensures consistent product quality. The insights obtained from this study will enable future progress in industrial automation, promoting continuous improvements and competitiveness in manufacturing industries.

References

1. Jovane F, Koren Y, Boer C (2003) Present and future of flexible automation: towards new paradigms. CIRP Ann 52(2):543–560
2. Kumar A, Parveen, Liu Y, Kumar R (eds) (2024) Handbook of intelligent and sustainable manufacturing: tools, principles, and strategies, 1st edn. CRC Press. https://doi.org/10.1201/9781003405870
3. Kumar A, Dogra N, Sarita, Bhatia S, Sidhu MS (eds) (2024) Handbook of intelligent and sustainable smart dentistry: nature and bio-inspired approaches, processes, materials, and manufacturing, 1st edn. CRC Press. https://doi.org/10.1201/9781003404934
4. Kumar A, Rajak DK, Kumar P, Rathee S (eds) (2024) Smart electric and hybrid vehicles: design, modeling, and assessment by industry 4.0 approaches, 1st ed. CRC Press. https://doi.org/10.1201/9781003502470
5. Kumar A, Rajak DK, Kumar P, Rathee S (eds) (2024) Smart electric and hybrid vehicles: fundamentals, strategies and applications, 1st ed. CRC Press. https://doi.org/10.1201/9781003495574
6. Basu R, Das MC, Kumar A, Sarkar B (2024) Identifying and prioritizing quality 4.0 practices in sustainable manufacturing using rough number-based AHP-MABAC. In: Handbook of intelligent and sustainable manufacturing: tools, principles, and strategies, 1st edn. CRC Press eBooks, pp 135–165. https://doi.org/10.1201/9781003405870-8
7. Kumar A (2024) Critical state-of-the-art literature review of surface roughness in incremental sheet forming: a comparative analysis. Appl Surf Sci Adv 23:100625. https://doi.org/10.1016/j.apsadv.2024.100625
8. Javaid M, Haleem A, Singh RP, Suman R (2021) Substantial capabilities of robotics in enhancing industry 4.0 implementation. Cogn Robot 1:58–75
9. Ren S, Zhang Y, Liu Y, Sakao T, Huisingh D, Almeida CM (2019) A comprehensive review of big data analytics throughout product lifecycle to support sustainable smart manufacturing: a framework, challenges and future research directions. J Clean Prod 210:1343–1365
10. Yaseen A (2021) Reducing industrial risk with AI and automation. Int J Intell Autom Comput 4(1):60–80
11. Park S-C (2018) The fourth industrial revolution and implications for innovative cluster policies. AI Soc 33:433–445
12. Messler RW (2004) Joining of materials and structures: from pragmatic process to enabling technology. Butterworth-Heinemann
13. Dasari S (2012) Damping of structures with riveted joints. Master Thesis, National Institute of Technology Rourkela, Rourkela, India
14. Xi F-F, Yu L, Tu X-W (2013) Framework on robotic percussive riveting for aircraft assembly automation. Adv Manuf 1:112–122
15. Sontakke K, Vaidya R, Mate D (2014) Design and analysis of drilling cum riveting machine. J Emerg Technol Innov Res 1:419–426
16. Xi F, Lin Y, Li Y (2017) A robotic percussive riveting system for aircraft assembly automation. Adv Mechatron MEMS Devices II:443–468
17. Pina Cipriano G (2019) Energy efficiency based analysis and optimization of friction riveting
18. Bagherpour E, Huang Y, Fan Z (2019) Assessment of self-piercing riveted joints using the analytic hierarchy process. Metals 9(7):760
19. Zdravkovic S, Golovashchenko S (2020) Analysis of sheet metal joining with self-piercing riveting. In: SAE Technical Paper, pp 0148–7191
20. Ribas Matas C (2021) Comparison of the response of different rivet layout patterns in aircraft repair patches. Universitat Politècnica de Catalunya
21. Das H, Kappagantula KS, Srivastava A, Upadhyay P, Dos Santos JF, Reza-E-Rabby M (2023) Embedded anchoring of multi-material assemblies by friction-riveting process. TMS Annual Meeting & Exhibition. Springer, pp 149–156

Real-World Security-Constrained AC Optimal Power Flow Analysis Using Artificial Gorilla Troops Optimizer

Sundaram B. Pandya, Kanak Kalita, Ajay Kumar, Pradeep Jangir, Shankar Chakraborty, and Laith Abualigah

Abstract The problem of Alternating Current Optimal Power Flow (ACOPF) pertains to the strategic identification of the most advantageous operating conditions for electricity generating facilities. The objective here is to fulfill the demand across a power transmission network while simultaneously curbing operating costs.

S. B. Pandya
Department of Electrical Engineering, Shri K.J. Polytechnic, Bharuch 392001, India

K. Kalita
Department of Mechanical Engineering, Vel Tech Rangarajan Dr. Sagunthala R&D Institute of Science and Technology, Avadi, Chennai 600062, India

University Centre for Research and Development, Chandigarh University, Mohali 140413, India

K. Kalita
e-mail: drkanakkalita@veltech.edu.in

A. Kumar (✉)
Department of Mechanical Engineering, School of Engineering and Technology, JECRC University, Jaipur, Rajasthan, India
e-mail: ajay.kumar30886@gmail.com

P. Jangir
Department of Biosciences, Saveetha School of Engineering, Saveetha Institute of Medical and Technical Sciences, Chennai 602105, India

Jadara Research Center, Jadara University, Irbid 21110, Jordan

S. Chakraborty
Production Engineering Department, Jadavpur University, Kolkata, India

L. Abualigah
Computer Science Department, Al Al-Bayt University, Mafraq 25113, Jordan

MEU Research Unit, Middle East University, Amman 11831, Jordan

Applied Science Research Center, Applied Science Private University, Amman 11931, Jordan

Centre for Research Impact and Outcome, Chitkara University Institute of Engineering and Technology, Chitkara University, Rajpura, Punjab 140401, India

Artificial Intelligence and Sensing Technologies (AIST) Research Center, University of Tabuk, Tabuk 71491, Saudi Arabia

A. Kumar et al. (eds.), *Trends in Mechatronics Systems*, Emerging Trends in Mechatronics, https://doi.org/10.1007/978-981-97-9108-8_2

The significance of the Optimal Power Flow (OPF) challenge has rapidly escalated in recent years, impacting the financial operations of modern power plants. The complexity of the OPF issue lies in its non-convex and high-dimensional nature. This makes it an intricate optimization challenge that requires the application of sophisticated metaheuristic algorithms for its resolution. The focus of this paper is on the utilization of an advanced algorithm, the Artificial Gorilla Troops Optimizer (AGTO), for scrutinizing some of the benchmark functions and addressing the real-world ACOPF problem. To ensure rigorous testing and validation, the study takes into account the IEEE 30-bus test system. We then compare our simulation results with other renowned optimization methodologies available in the existing literature. This comparative study demonstrates that the application of AGTO could potentially yield high-quality OPF solutions, exhibiting its potency in addressing the ACOPF problem.

Keywords Optimal power flow · AGTO optimizer · ACOPF problem · Metaheuristic algorithms · Power system optimization · Renewable energy

1 Introduction

Corrective power flow analysis occupies a critical role in the electrical network planning process, ensuring the security and efficiency of power system operations. Optimal Power Flow (OPF) approach, which is one of the significant tools of electrical engineering, plays a crucial role in determining feasible operating points by minimizing or maximizing certain objectives subject to a set of constraints. This approach not only guarantees satisfaction of multiple equality and inequality constraints but also meets the requirements of practical operation in power systems. The history of OPF can be regarded to have started from the work done by Carpentier in 1962 [1] when the world was introduced to the concept of optimizing power systems.

Since then, OPF has evolved into a versatile field of study that is vital to the scheduling and operational planning of the modern power systems. Its importance has been on the rise with the development of complex electrical systems that need proper tools for analysis and decision making. In Niu et al. [2] elaborate on this path of development and note that a number of stochastic methods have been considered for the OPF problem, especially within the last few years. Despite the potential of metaheuristic algorithms in some facets of OPF, the algorithms' efficiency differs based on problem sets since some algorithms may not work well when applied in different situations [3].

The past decade has witnessed a proliferation of methodologies aimed at solving the OPF problem, demonstrating the vibrant and dynamic nature of this research area. The comprehensive review by Niu et al. [2] provides an invaluable resource, summarizing various metaheuristic algorithms that have been employed to solve OPF challenges, complete with a detailed list of these algorithms and their abbreviations as presented in Table 1. In response to this rich history and the ongoing development of OPF solutions, our paper engages with several of these metaheuristic approaches, conducting a comparative analysis to explore their efficacy in the context of our case studies.

At its core, OPF represents a fundamental optimization challenge within power systems engineering, aimed at the efficient scheduling and dispatch of power-generating units. By striving to minimize total generation cost and line losses, while satisfying a comprehensive set of operational and physical constraints, OPF facilitates the determination of optimal operating conditions. This, in turn, plays a crucial role in the planning, operation, and market activities associated with power systems, significantly impacting their economic and secure operation. Through our investigation, we aim to contribute to the rich tapestry of OPF research, offering insights into the application and potential of advanced optimization algorithms in meeting the evolving needs of modern power systems.

The current developments in power system optimization have concentrated on the incorporation of renewable energy sources together with the traditional power plants. This change is due to the need to increase productivity and efficiency of electrical systems production. For instance, the analysis that combines coal and wind-based power units has employed complex optimization techniques such as the Gbest guided Artificial Bee Colony (GABC) algorithm [33] to enhance the results of the OPF. This approach not only helps in dealing with the inherent variability of wind power but also aims at integrating the output of various energy sources.

A tool like the metaheuristic algorithms Ant Colony Optimization (ACO) [34] and Biogeography-Based Algorithm (MBFA) [35] are part of a trend in solving OPF problem from different computational perspectives. For further helping in the coordination of VAR power production, especially with technologies like Endesa and Iberdrola working on Different Feed Induction Generators (DFIG) and Static Synchronous Compensators (STATCOM), respectively. These technologies involve more efficient power distribution which offer the possibility of dynamic Power System management and integration with high penetration level with renewable energy sources [36].

The OPF problem is non-convex and high-dimensional, posing formidable analytical challenges. The situation is even more difficult to handle when considering multiple objectives that are in conflict with each other, a large number of variables and several operational constraints together [37, 38]. The increasing penetration of renewable energy sources further complicates this picture, meaning that robust and efficient optimization algorithms are needed to navigate these complexities. Traditional methods like gradient based, Newton–Raphson end up being cumbersome and slow to converge but Riley & Williamson in [39] further explain that some new classes of algorithms have been developed e.g. optimal discretization techniques or model order reduction which are difficult management (non-linear, non-convex problems).

More and more, sophisticated OPF scheduling algorithms are considering the incorporation of renewable resources to improve grid sustainability and robustness. Tazvinga et al. A scheduling strategy is designed for a hybrid network, made up by solar photovoltaic (SPV) systems and multi-storage devices of battery energy

Table 1 Summary of different metaheuristic algorithms used for OPF problem

Algorithm	Description	Sources	Ref
ABC	Artificial Bee Colony	Kılıç and Ayan, 2013	[3]
BBO	Biogeography Based Algorithm	Dutta et al., 2015	[4]
BHBO	Black Hole Based Optimization	Bouchekara, 2014	[5]
CPSO	Chaotic Particle Swarm Optimization	Khaled et al., 2017	[6]
DE	Differential Evolution	Sayah and Hamouda, 2019	[7]
DSA	Differential Search Algorithm	Bouchekara and Abido, 2014	[8]
EADDE	Evolving Ant Direction Differential Evaluation	Vaisakh et al., 2013	[9]
EADHDE	Genetic Evolving Ant Direction HDE	Vaisakh and Srinivas, 2011	[10]
EGA	Enhanced Genetic Algorithm	Bakirtzis et al., 2002	[11]
EP	Evolutionary Programming	Yuryevich and Wong, 1999	[12]
FGA	Fuzzy Genetic Algorithm	Mahdad et al., 2008	[13]
FPSO	Fuzzy Particle Swarm Optimization	Khaled et al., 2017	[6]
GA	Genetic Algorithm	Leung and Chung, 2000	[14]
GM	Gradient Method	Woo et al., 2020	[15]
GSA	Gravitational Search Algorithm	Duman et al., 2012	[16]
GSO	Group Search Optimization	Tan et al., 2015	[17]
GTOT	Gorilla Troops Optimization Technique	Shaheen et al., 2022	[18]
GWO	Gray Wolf Optimization	Hassan and Zellagui, 2018	[19]
HBOA	Heap-Based Optimization Algorithm	Shaheen et al., 2022	[20]
HS	Harmony Search	Sinsuphan et al., 2013	[21]
ICA	Imperialistic Competitive Algorithm	Hmida et al., 2019	[22]
MRFO	Manta Ray Foraging Optimizer	Kahraman et al., 2022	[23]
MSA	Moth Swarm Algorithm	Mohamed et al., 2017	[24]
MSFLA	Modified Shuffle Frog Leaping Algorithm	Niknam et al., 2011	[25]
MVO	Multi-Verse Optimizer	Bentouati et al., 2016	[26]
PSO	Particle Swarm Optimization	Khaled et al., 2017	[6]
SCA	Sine–Cosine Algorithm	Weng et al., 2023	[27]
SFLA	Shuffle Frog Leaping Algorithm	Rahiminejad et al., 2014	[28]
SMA	Slime Mold Algorithm	Farhat et al., 2022	[29]
TLBO	Teaching–Learning-Based Optimization	Alanazi et al., 2022	[30]
TS	Tabu Search	Ongsakul and Bhasaputra, 2002	[31]
TSA	Tunicate Swarm Algorithm	Logeswaran et al., 2022	[32]

storage system-BESSs work together with diesel generators in [40]. Pumped hydro storage was proposed by them as an alternative for energy storage, highlighting hybrid systems sturdiness and flexibility in erratic changes of energy demands/supply [41].

The complexity of managing diverse energy sources in OPF problems necessitates sophisticated design and optimization strategies. According to [42, 43], a comprehensive design that incorporates solar, wind, and hydro resources alongside dynamic economic dispatch could significantly improve power system efficiency and reliability [44]. Salkuti [45] demonstrated the potential of combining coal-based power with solar and wind energy, further augmented by battery storage, to meet varying energy demands effectively.

Recent advancements in algorithmic solutions have shown promising results in tackling the OPF problem's inherent complexities. The use of the non-dominated sorting genetic algorithm-II (NSGA-II) was proposed as an effective tool for managing modern electrical systems with multiple objectives, showcasing superior performance in simulation studies [46, 47]. Furthermore, Biswas et al. [48] introduced a novel approach to evaluating OPF problems through variable probability density functions for renewable energy sources, offering a more nuanced analysis of potential outcomes.

The integration of renewable energy sources introduces significant variability and uncertainty into power system operations. This challenge is compounded in hybrid AC/DC networks, where the optimal placement of VSC-HVDC converters is critical for system efficiency and stability [49]. The need for advanced optimization algorithms is more pressing than ever, as traditional methods like gradient-based and Newton–Raphson algorithms struggle with the nonlinear and non-convex nature of OPF problems.

Efficient and robust optimization algorithms are essential for navigating the complexities of the OPF problem, especially as the integration of renewable energy sources and demand response mechanisms continue to grow. These algorithms must be capable of handling multiple conflicting objectives, a vast array of variables, and stringent operational constraints, all while ensuring system security and resilience against contingencies.

The use of metaheuristic algorithms based on animal behavior has been a popular area of research, and many algorithms are based on the complex interactions between predators and their prey. However, all these algorithms are diverse and none of them has attempted to capture the social organization and interactions of a gorilla troop. To address this issue, this research proposes an innovative algorithm based on these social processes, referred to as the Artificial Gorilla Troops Optimizer (AGTO) [50]. This new class of swarm-intelligence-based algorithm is distinctive because it presents a new approach that can be associated with the behavior of gorillas in order to improve the performance of certain optimization mechanisms.

The development of AGTO was inspired by the absence of heuristics that exploit the special features of gorilla culture, which is based on close-knit relationships and rational thinking. It has been used in different fields such as mechanical design optimization, scheduling problems, and network routing and has proved to be efficient and reliable. In its design, it has both exploration and exploitation modes which mimic the actions of a gorilla troop, including foraging and group choices. These features make AGTO especially suitable for-searching the solution space of optimization problems while providing a balance between the exhaustive and efficient approaches.

Even though AGTO has shown promising results in a few areas, its utilization in solving the particular challenges of security-constrained AC OPF problems is still limited. The challenges that arise from these problems include their non-convexity and the requisite consideration of security constraints. This research intends to fill this gap by investigating the AGTO approach on the AC OPF problem, a significant issue in the domain of power system optimization. In this work, we are interested in enhancing the understanding of AGTO's functionalities and applying it to address the more complex issues of real-world power systems.

The remainder of this paper is organized as follows: Sect. 2 frames the OPF issue; Sect. 3 provides a brief discussion on the proposed AGTO algorithm; Sect. 4 delves into the applications and results of the real-world ACOPF; and finally, Sect. 5 consolidates our findings and conclusions.

2 Optimum Power Flow Problem Formulation

The generalized OPF problem is represented as per Eq. (1). The OPF problem is tasked with identifying the optimal state of the power system, achieved through an optimal control setting. This is done by realizing a defined objective function value that simultaneously fulfills operational constraints. In essence, the goal of OPF is to strike a balance between efficiency and reliability, ensuring optimal power delivery while adhering to all necessary system constraints.

$$
\begin{aligned}
&\min f(x, u) \\
&\text{S.t } g(x, u) = 0 \\
&h(x, u) \leq 0
\end{aligned}
\tag{1}
$$

where $f(x, u)$ defines an objective function either single or multi-objective. Term g and h are the operational equality and inequalityconstraints respectively. Terms x and u represent thesystem state and control variables. Four control variables are considered in this analysis are:

(I) PG: Generator active power output except at slack bus
(II) VG: Generator bus voltage;
(III) T: Online transformer tap setting
(IV) QC: Reactive power compensation.

The control set vector is expressed as below,

$$
u^T = \left[p_{G_2} \ldots p_{G_{NG}}, V_{G_1} \ldots V_{G_{NG}}, T_1 \ldots T_{NT}, Q_{C_1} \ldots Q_{C_{NC}}\right] \tag{2}
$$

Similarly, four state variables considered in this analysis includes,

(I) PG1: Active power output, generated by slack bus;
(II) VL: voltage generated by load buses;
(III) QG: Reactive power output of generator;
(IV) SL: Line loading capacity of transmission lines.

Equation (3) represents the state variable vector.

$$x^T = \left[p_{G_1}, V_{L_1} \ldots V_{L_{NL}}, Q_{G_1} \ldots Q_{G_{NG}}, S_{l_1} \ldots S_{l_{nl}}\right] \tag{3}$$

In above expressions symbols *NG*, *NT*, *NC*, *NL*, and *nl* are respectively numbers of generators, tap-changing transformers, Shunt compensators, load buses and transmission lines.

2.1 Objective Functions

OPF studies analyzes following sets of objective functions, explained in brief as follows,

2.1.1 Total Generation Cost (TGC) Minimization

The first single objective function considering of the minimization of TGC of power station calculated based on quadratic curve in Eq. (4).

$$F1 = C(P) = \sum_{i=1}^{NG} \left(a_i P_{gi}^2 + b_i P_{gi} + c_i\right) \text{ in } \frac{\$}{\text{Hr}} \tag{4}$$

where a_i, b_i and c_i are the cost constants of i th generator.

2.1.2 Emission Minimization

This function measures the amount of pollutant due to carbon emission to identify environmental impact and aimed to minimize. Emission is calculated as follows,

$$F2 = \sum_{i=1}^{NG} (\alpha_i + \beta_i P_{gi} + \gamma_i P_{gi}^2 + \xi_i \exp(\lambda_i P_{gi}) \text{ in Ton/Hr.} \tag{5}$$

where γ_i, β_i, α_i, ξ_i and λ_i are the emission constants of i th unit.

2.1.3 Active Power Loss Minimization

The total transmission active power loss is desire to minimize and formulated as per Eq. (6).

$$F3 = P_{\mathrm{LOSS}} = \sum_{i=1}^{NB} P_{Gi} - \sum_{i=1}^{NB} P_{Di} \text{ in } Mw \tag{6}$$

where NB is the number of buses; P_{Gi} and P_{Di} are generation and dispatch of i th bus.

2.1.4 Reactive Power Loss Minimization

The static voltage stability margin is inherently dependent on the availability of reactive power support for power transmission from one endpoint to another. In simpler terms, it is crucial for a steady supply of reactive power to ensure the uninterrupted and efficient transmission of electricity across the network. Thus, the loss of reactive power is determined by the following equation, reflecting the direct relationship between static voltage stability margin and reactive power support.

$$F4 = Q_{\mathrm{LOSS}} = \sum_{i=1}^{NB} Q_{Gi} - \sum_{i=1}^{NB} Q_{Di} \text{ in } M\,\text{var} \tag{7}$$

2.1.5 Voltage Deviation Minimization

Voltage deviation quantifies the disparity in the magnitude of bus voltage when compared to a standard value of 1.0 per unit (pu). It is critical to maintain the minimum required value of voltage deviation to ensure the secure operation of buses within the electrical system. The corresponding objective function, which conveys this concept, can be expressed as follows:

$$F5 = \sum_{i=1}^{N_{pq}} |v_i - 1.0| \text{ in } p.u \tag{8}$$

where N_{pq} the number of load buses, v_i is the voltage level of i th bus in p.u.

2.1.6 Voltage Steadiness Improvement (L-index)

The L-index denotes the voltage stability margin of each bus, which aids in maintaining a steady voltage within an appropriate range under standard operating conditions. It provides a scalar value to each load bus, representing its voltage stability. The L-index ranges from zero, indicating a no-load case, to one, signifying a voltage collapse. In essence, the L-index serves as a measure or indicator of the severity of potential voltage failure of j th bus can be obtained as:

$$L_j = \left| 1 - \sum_{i=1}^{N_g} F_{ji} \frac{V_i}{V_j} \right| \forall j = 1, 2, \ldots\ldots, NL \tag{9}$$

$$F_{ji} = -[Y_1]^{-1}[Y_2] \tag{10}$$

where Y_1 and Y_2 are the sub-matrices of Y_{BUS}. The role of voltage stability improvement can be expressed as

$$F6 = L = \max(L_j) \forall j = 1, 2, \ldots\ldots, NL \tag{11}$$

2.2 *Constraints*

2.2.1 Equality Constraints

Static load flow equations constitute the equality constraints set and represented in Eqs. (12) and (13).

$$P_{Gi} - P_{Di} - V_i \sum_{j=1}^{N} v_j \left(G_{ij} \cos(\delta_i - \delta_j) + B_{ij} \sin(\delta_i - \delta_j)\right) = 0; i = 1, 2, 3, \ldots\ldots N \tag{12}$$

$$Q_{Gi} - Q_{Di} - V_i \sum_{j=1}^{N} v_j \left(G_{ij} \sin(\delta_i - \delta_j) + B_{ij} \cos(\delta_i - \delta_j)\right) = 0; i = 1, 2, 3, \ldots\ldots N \tag{13}$$

Different symbols used in above equations represented following variables of power system network. N is the number of buses; P_{Gi} and P_{Di} are real power generation and demand; Q_{Gi} and Q_{Di} are reactive power generation and demand; i and j are the indices representing respectively i thand j thbus; V and δ are the bus voltage magnitude and angle. G_{ij} and B_{ij} are the real and imaginary parts of the ij th element of the bus admittance matrix, respectively.

2.2.2 Inequality Constraints

The inequality constraints $h(x,u)$ are very important which having the operating limits of the power system networks and given as below;

$$P_{Gi\min} \leq P_{Gi} \leq P_{Gi\max} i = 1, \ldots\ldots Ng \tag{14}$$

$$Q_{Gi\min} \leq Q_{Gi} \leq Q_{Gi\max} i = 1, \ldots\ldots Ng \tag{15}$$

$$V_{Gi\min} \leq V_{Gi} \leq V_{Gi\max} i = 1, \ldots\ldots Ng \tag{16}$$

$$T_{i\min} \leq T_i \leq T_{i\max} i = 1, \ldots\ldots Nt \tag{17}$$

$$Q_{ci\min} \leq Q_{ci} \leq Q_{ci\max} i = 1, \ldots\ldots Nc \tag{18}$$

$$V_{Li\min} \leq V_{Li} \leq V_{Li\max} i = 1, \ldots\ldots Npq \tag{19}$$

$$|S_{Li}| \leq S_{Li\max} i = 1, \ldots\ldots\ldots\ldots N_l \tag{20}$$

In above equations, Eqs. (14–16) form generator constraints set, Eq. (17) form transformer constraints set; Eq. (18) form VAR resource constraints set and Eqs. (19) and (20) are the security constraints sets.

3 Artificial Gorilla Troops Optimizer (AGTO) Algorithm

In 2020, Benyamin and his team introduced the Gorilla Troops Optimizer (AGTO), a metaheuristic algorithm inspired by the behavior of gorilla troops, as illustrated in Fig. 1. In AGTO, the troop's social structure is abstracted into three components: silverback (the leader), blackbacks (the followers), and juveniles (the explorers). The algorithm starts by generating a random population of solutions, each represented by a gorilla. The fitness of each solution is evaluated, and the solution with the highest fitness is designated as the silverback. The search for the optimal solution is carried out through three primary operations, which mirror gorilla behaviors: migration (the search for new habitats), fighting (competition for leadership), and collaboration (collective problem-solving). Migration helps explore the search space, while fighting encourages exploitation of the current best solution. Collaboration aids in

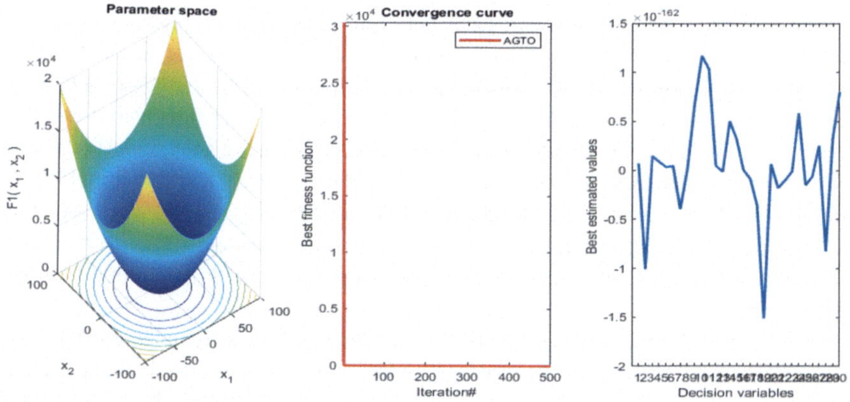

Fig. 1 Qualitative results for the F1 with 30 dimensions

balancing exploration and exploitation by considering the input from both the silverback and other gorillas. The solution with the best fitness at the end of the iterations is taken as the optimal solution.

This algorithm integrates a variety of optimization procedures, showcasing enhanced robustness in problem-solving and remarkable search capabilities across a wide spectrum of optimization spaces. According to experimental results, the AGTO offers superior convergence and accuracy compared to most existing optimization algorithms, making it well-suited for real-world case studies featuring undefined search spaces. The AGTO comprises two main phases: the exploration phase and the exploitation phase. The exploration phase involves three distinct mechanisms. The first mechanism entails relocating to an unknown area, which enhances the algorithm's comprehensive surveillance of the entire problem space. The second mechanism involves moving to a familiar location, which improves the algorithm's detection capabilities. The final mechanism promotes migration among other gorillas, thereby enhancing the algorithm's ability to avoid local optima. The respective simulation equations for these three mechanisms are detailed below:

$$GX(t+1) = \begin{cases} (UB - LB) \times r_1 + LB, & rand < p, \\ (r_2 - C) \times X_r(t) + L \times H, & rand \geq 0.5, \\ X(i) - L \times ((L + r_3) \times (X(t) - GX_r(t))), & rand < 0.5, \end{cases} \tag{21}$$

$$C = [\cos(2r_4) + 1] \times \left(1 - \frac{I_t}{\max I_t}\right) \tag{22}$$

$$L = C \times l \tag{23}$$

$$H = Z \times X(t) \tag{24}$$

The variables are updated with each iteration using Eq. (21). Here, $GX(t+1)$ and $X(t)$ are respectively candidate position of the gorilla vector in successeive iteration and gorilla's current position vector. Different random numbers sets with their range are $r_1, r_2, r_3, r_4, rand$ [0, 1]; l [−1, 1]; Z, problem dimension; and the range is $[-C, C], p,$, which defines the likelihood of choosing an unidentified place for migration with a range of [0,1], that must be set in advance. The higher and lesser boundaries of the variable are denoted by *UB* and *LB*, respectively, while X_r and GX_r indicate the location trajectory of a gorilla member randomly chosen from the relevant population, I_t is the current iteration count, and $\max I_t$ is the total number of iterations.

Following the silver-backed gorilla and vying for mature females are the two strategies used in the exploitation phase, and they are chosen based on the values of C and W. Equation 26 is used to simulate this behavior when W is the preset parameter and the former method is used if $C \geq W$.

$$GX(t+1) = L \times \left(\left| \frac{1}{N} \sum_{i=1}^{N} GX_i(t) \right|^{2^L} \right)^{\frac{1}{2^L}} \times (X(t) - X_{\text{silverback}}) + X(t) \tag{25}$$

where $X_{\text{silverback}}$ is the silverback gorilla locationtrajectory, viz. the best result.

If $C < W$ then select the next mechanism and put on this performance with Eq. (28).

$$GX(i) = X_{\text{silverback}} - \left[X_{\text{silverback}} Q - X(t) Q\right] \beta E \tag{26}$$

$$Q = 2 \times r_5 - 1 \tag{27}$$

$$E = \begin{cases} N_1, rand \geq 0.5 \\ N_2, rand < 0.5 \end{cases} \tag{28}$$

The random numbers between [0, 1] are r_5 and *rand*. The effects of the conflict are mimicked using Q. β is a predetermined stricture. E is used to pretend the effect of strength on the measurement of the answer. The value of E is equivalent to the arbitrary value in the standard distribution and the problem's dimension if $rand \geq 0.5$; otherwise, it is equal to the random value in the normal distribution. Conduct a group formation operation to determine the price of GX at the conclusion of this phase. Identify the site of the silverback gorilla in the population, which is the best option, if $GX(t) < X(t)$, and then choose $GX(t)$ as the solution of $X(t)$. For the application of AGTO to the OPF problem, some modifications are required to handle the unique characteristics of the problem. The objective function and constraints of the OPF problem are integrated into the fitness evaluation of the AGTO. Each gorilla (solution) is a vector representing generator outputs, and its fitness is determined by the total generation cost or line loss, subject to operational and security constraints.

Also, to handle the non-linear and non-convex nature of the OPF problem, a suitable encoding and decoding mechanism is implemented. Problem-specific operators can be introduced to improve the efficiency of the migration, fighting, and collaboration operations.

4 Results and Discussion

4.1 Benchmark Set and Compared Algorithms

In order to affirm the effectiveness and efficiency of the proposed algorithm, we've leveraged a system of benchmark optimization models as our testing apparatus. The numerical experiments were conducted on a computer with a Core i5 processor operating at 3.20GHz, equipped with 4GB of RAM. The AGTO algorithm was developed using the MATLAB programming language. To investigate the performance of AGTO, we evaluated it alongside other optimization methodologies. To gauge AGTO's scalability, we utilized a series of benchmark functions, extending from F1 through F13. The results, depicted in Figure 1, suggest that AGTO harbors significant potential to yield satisfactory outcomes and performance across all dimensions. As the dimensions increased, AGTO managed to maintain an appropriate level of search capabilities. Contrarily, the performance of other comparable optimizers saw a noticeable decline with the addition of dimensions, giving AGTO a considerable advantage across all comparable dimensions in the F1–F13 range. Scalability testing distinctly reveals that AGTO excels at achieving the ideal balance between exploration and exploitation competencies when confronted with substantial challenges. Table 2 presents the parameters for the optimization techniques employed in this comparative evaluation of AGTO. It also portrays the best parameter setting adopted by various algorithms. The results in Table 3 illustrate that with the selected parameters for AGTO, the derived outcomes surpass those obtained with other metaheuristic algorithms.

Table 2 Parameters for optimization algorithms

Algorithms	Parameters	Values
AGTO	β W p	3 0.8 0.03
TSA	Parameter P_{max} Parameter P_{min}	4 1
GWO	Convergence Constant α	[2 0]
SCA	A	2
MVO	Existence probability Traveling distance rate	[0.2 1] [0.6 1]

Table 3 Outcomes of standard functions (F1-F13), with 30 dimensions

Function	AGTO	TSA	GWO	SCA	MVO
F1	**0**	1.4368E − 23	3.3533E − 29	3.9005E − 03	8.3647E − 01
F2	**1.04E-199**	5.3213E − 15	1.3643E − 17	2.7766E − 04	5.4848E − 01
F3	**0**	4.9700E − 09	3.8126E − 08	6.69E + 02	6.51E + 01
F4	**5.85E-201**	1.5499E − 02	8.1138E − 08	1.17E + 01	1.18E + 00
F5	**5.63E-05**	2.62E + 01	2.58E + 01	4.94E + 01	3.84E + 01
F6	**6.82E-09**	1.80E + 00	7.7955E − 05	4.13E + 00	8.0166E − 01
F7	**2.73E-05**	2.7697E − 03	3.5203E − 04	3.6638E − 03	1.6213E − 02
F8	**-12,569.48662**	− 7.3162E + 03	− 7.2007E + 03	− 4.4667E + 03	− 9.5122E + 03
F9	**0**	1.03E + 02	5.6843E − 14	3.1313E − 03	8.90E + 01
F10	**8.88E-16**	1.3580E − 12	7.1942E − 14	1.3046E − 01	1.13E + 00
F11	**0**	0.00E + 00	0.00E + 00	3.0943E − 01	7.9821E − 01
F12	**8.87E-12**	3.4138E − 01	1.4121E − 02	9.1971E − 01	6.7568E − 02
F13	**2.20E-10**	1.88E + 00	2.2815E − 01	2.51E + 00	3.8911E − 02

This figure provides a visual representation of the AGTO's social hierarchy, including silverbacks (leaders), blackbacks (followers), and juveniles (explorers). This structure underpins the algorithm's dynamics, where different roles contribute to the exploration and exploitation phases in the optimization process. This figure provides a visual representation of the AGTO's social hierarchy, including silverbacks (leaders), blackbacks (followers), and juveniles (explorers). This structure underpins the algorithm's dynamics, where different roles contribute to the exploration and exploitation phases in the optimization process. Depicts the strategies employed in the exploitation phase, focusing on following the silverback and competing for leadership. These mechanisms ensure intensive exploitation around promising regions identified during the exploration phase, fine-tuning the search toward optimal solutions.

4.2 *Real-World AC Optimum Power Flow Problem*

The proposed approach has been demonstrated using the IEEE-30 bus system. The active and reactive demands for this test bus system stand at 2.834 per unit (p.u.) and 1.262 p.u., respectively, considering a base of 100 MVA. The load bus voltage ranges between 0.95 and 1.1 in p.u., representing the minimum and maximum points, respectively. In addition, transformer tap positions with constraints spanning between 0.9 and 1.1 p.u. are taken into account. The system also includes adjustable shunts with limitations between 0 and 5 MVAr. We've considered the main characteristics of the IEEE-30 bus system as detailed in the works by Pandya and Jariwala [46,

47] for this proposed analysis. The AGTO is used to optimize all the objective functions mentioned in the problem formulation, treating them as individual objective optimization problems. The optimum minimum values for these objective functions for each scenario, spanning from 1 to 6, are tabulated in Table 4, along with their corresponding control variable values.

Table 4 Outcomes of IEEE 30 bus scheme with AGTO algorithm

Control variables	Case-1	Case-2	Case-3	Case-4	Case-5	Case-6
P_{G1}	200.00	159.96	200.00	200.00	50.05	200.00
P_{G2}	49.28	67.34	80.00	80.00	80.00	80.00
P_{G5}	21.29	50.00	50.00	50.00	15.05	50.00
P_{G8}	20.84	35.00	35.00	34.99	35.00	35.00
P_{G11}	11.87	30.00	30.00	30.00	10.00	30.00
P_{G13}	12.00	40.00	40.00	39.98	40.00	40.00
V_{G1}	1.10	1.10	1.10	1.10	0.95	1.10
V_{G2}	1.09	1.10	1.10	1.09	1.09	1.10
V_{G5}	1.06	1.08	1.10	1.07	1.01	1.10
V_{G8}	1.07	1.09	1.10	1.08	1.00	1.10
V_{G11}	1.10	1.10	1.10	1.10	1.03	1.10
V_{G13}	1.10	1.10	1.10	1.10	0.97	1.10
$T_{11(6-9)}$	0.90	1.07	1.10	1.10	0.90	1.10
$T_{12(6-10)}$	1.10	1.02	1.10	1.10	0.98	1.10
$T_{15(4-12)}$	0.97	0.92	1.10	1.10	0.91	1.10
$T_{36(28-27)}$	0.93	0.90	1.10	1.10	0.96	1.10
QC_{10}	5.00	5.00	5.00	5.00	0.00	5.00
QC_{12}	1.90	5.00	5.00	5.00	2.41	5.00
QC_{15}	3.79	5.00	5.00	5.00	5.00	5.00
QC_{17}	5.00	5.00	5.00	5.00	5.00	5.00
QC_{20}	5.00	5.00	5.00	5.00	5.00	5.00
QC_{21}	5.00	5.00	5.00	5.00	5.00	5.00
QC_{23}	5.00	3.24	2.90	3.23	5.00	5.00
QC_{24}	5.00	4.46	5.00	4.94	5.00	5.00
QC_{29}	3.29	2.90	5.00	5.00	2.63	5.00
FC	**799.14**	–	–	–	–	–
EMISS	–	**0.21**	–	–	–	–
Ploss	–	–	**2.98**	–	–	–
Qloss	–	–	–	**14.97**	–	–
VD	–	–	–	–	**0.12**	–
Lmax	–	–	–	–	–	**0.11**

For the IEEE 30-bus system, Case 1's total Transmission Generation Cost (TGC) amounts to 799.1363 dollars per hour, Case 2's greenhouse gas emission is at a rate of 0.205623 tons per hour, and Case 3's individual power line real loss stands at 2.97746 MW. To maintain the voltage stability across various buses, the system requires a certain amount of reactive power. In instance 4, this power loss is calculated as 14.96985 MVAr. Similarly, one of the essential factors in ensuring the power system's reliability is the degree of deviation of each bus voltage from 1.0 per unit. Consequently, in Case 5, the lowest recorded voltage deviation is 0.121795 p.u. The voltage stability index, also referred to as the L-index, spans a range from 0 (when the system carries no load) to 1 (when the system is fully loaded). Hence, for Case 6, the minimum value for the voltage stability index is identified as 0.111416. When compared to other optimization methodologies (referencing the works of Pandya and Jariwala [46, 47]), the results for single objectives are deemed superior, as illustrated in Fig. 2. The convergence properties for Cases 1 through 6 are depicted in relation to the number of iterations in Fig. 3.

Comparing the AGTO algorithm with other specified techniques for the simultaneous reduction of multiple conflicting objective functions, it becomes clear from Fig. 2 that the results of AGTO with Other Optimization Methodologies Fig. 2a: Total Transmission Generation Cost (TGC) Minimization—This graph illustrates the AGTO's efficiency in minimizing the total generation cost when compared with other algorithms. The AGTO shows a significant reduction in cost, highlighting its effectiveness for economic dispatch in power systems. Figure 2b: Emission Minimization—Displays the AGTO's capability in reducing greenhouse gas emissions from power generation. This scenario underscores the AGTO's potential in contributing to environmentally sustainable power system operations. Figure 2c: Active Power Loss Minimization—Demonstrates the AGTO's performance in minimizing the power loss in transmission lines. The findings of this study show that AGTO can help in curbing losses hence enhancing the performance of the power transmission network. Figure 2d: AGTO Reactive Power Loss Minimization—This deals with the capacity of the AGTO in minimizing the reactive power loss. Reactive power loss is also an important factor that affects the voltage stability and power flow in the network. Figure 2e: Voltage Deviation Minimization—This feature demonstrates the ability of the AGTO in reducing voltage deviation from the desired levels at the load buses to prevent frequent malfunction of electrical appliances and equipment. Figure 2f: Voltage Stability Index (L-index) Minimization—Shares how the AGTO improves the voltage stability of the system. The L-index results show that the AGTO has effectively regulated voltage within the normal and safe range as the load varied.

Figure 2 encapsulates the AGTO's comparative performance across different optimization scenarios within the IEEE 30-bus test system framework. The Fig. 2a–f collectively demonstrate the AGTO's versatility and robustness in addressing multiple aspects of the OPF problem, from economic efficiency to environmental sustainability and system stability.

To demonstrate the effectiveness and efficiency of the proposed AGTO, a detailed convergence analysis was performed for each of the six case studies using the IEEE 30-bus test system. The convergence characteristics for these case studies are depicted

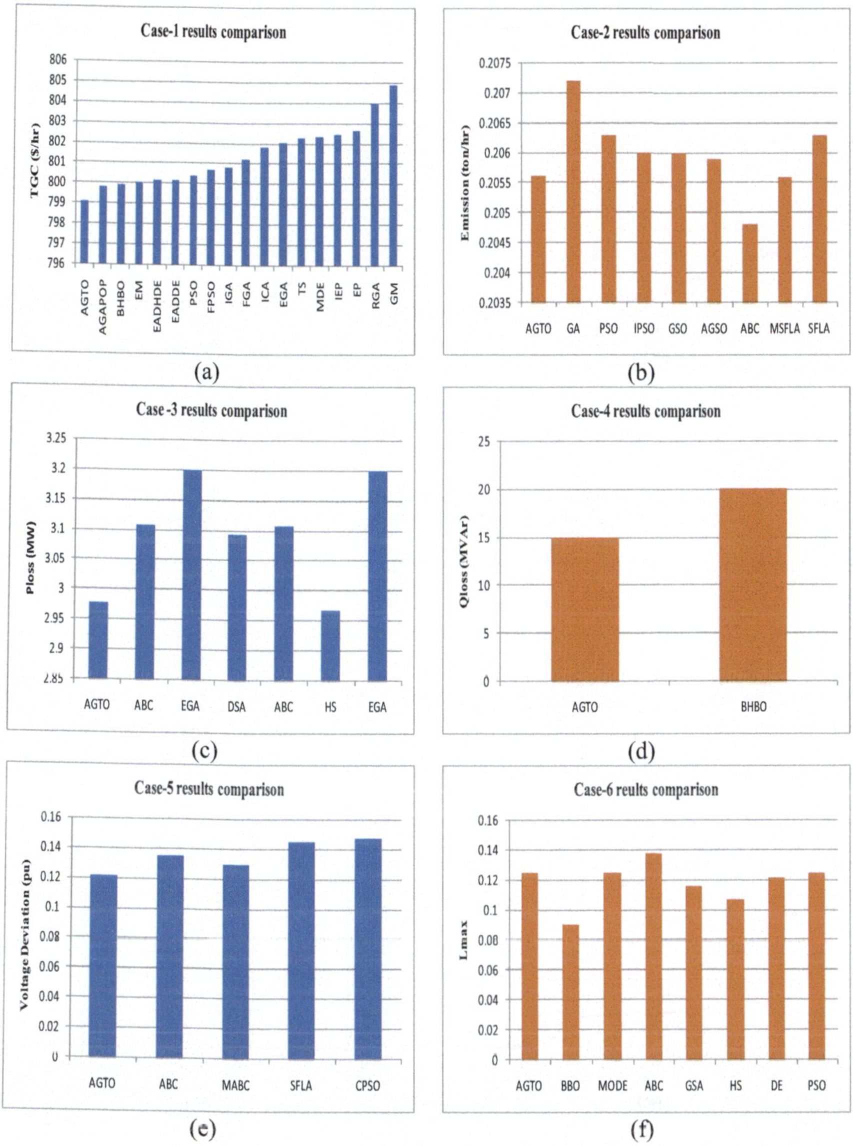

Fig. 2 Comparative results for **a** Case-1, **b** Case-2, **c** Case-3, **d** Case-4, **e** Case-5, and **f** Case-6

in Fig. 3, which is divided into Fig. 3a–f, representing the convergence properties of each scenario. Figure 3a: Total Transmission Generation Cost (TGC) Minimization Convergence Characteristics—This graph shows the convergence of AGTO in minimizing the total generation cost compared to other algorithms. The steep slope indicates AGTO's rapid achievement of optimal or near-optimal solutions, highlighting

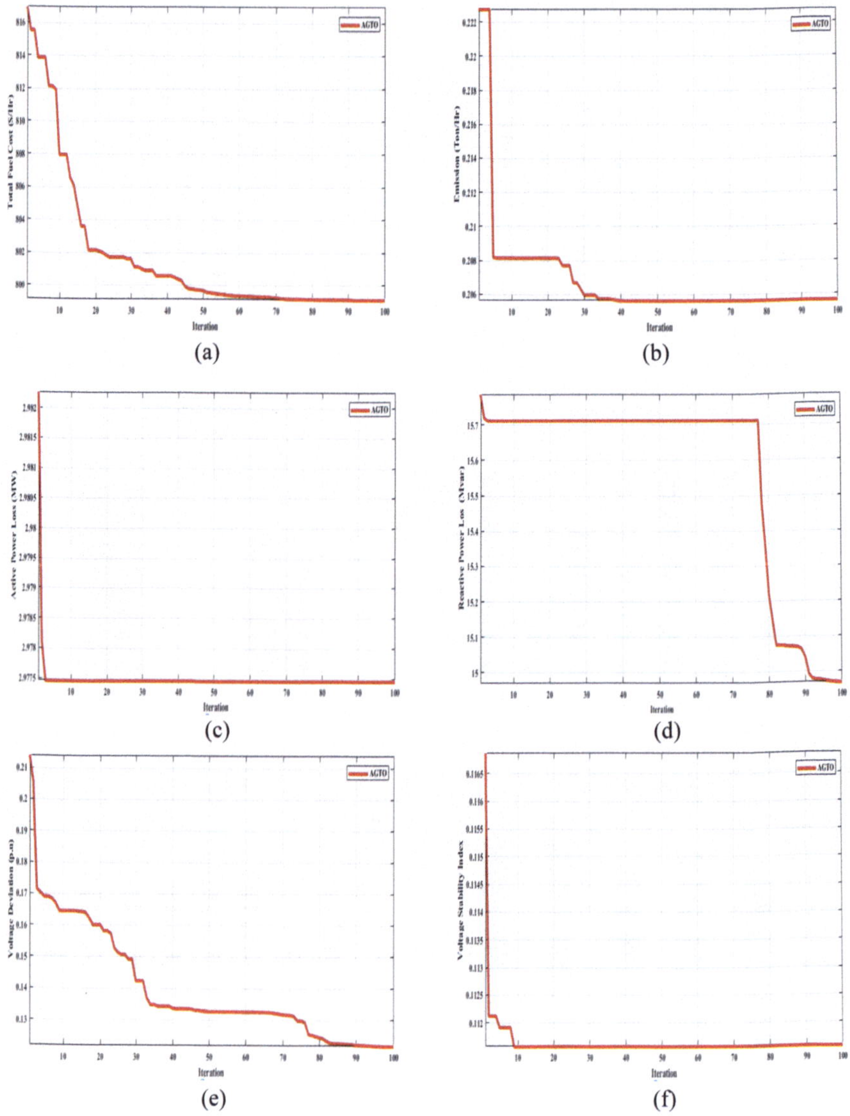

Fig. 3 **a** TGC minimization conversion characteristics, **b** emission minimization convergence characteristics, **c** active power loss minimization convergence characteristics, **d** reactive power loss minimization convergence characteristics, **e** voltage deviation minimization convergence characteristics, and **f** voltage stability index minimization convergence characteristics

its efficiency in economic dispatch optimization. Figure 3b: Emission Minimization Convergence Characteristics—Illustrates how AGTO approaches the minimum emissions goal. The graph underscores the algorithm's capability to swiftly converge toward solutions that significantly reduce greenhouse gas emissions from power generation. Figure 3c: Active Power Loss Minimization Convergence Characteristics—Demonstrates AGTO's convergence behavior in reducing transmission line losses. The curve reflects AGTO's ability to efficiently find solutions that enhance the power system's operational efficiency by minimizing power losses. Figure 3d: Reactive Power Loss Minimization Convergence Characteristics—Provides an insight of AGTO's capability in minimizing the reactive power loss. This is important in the regulation of voltage and efficient electricity distribution, and the graph shows the capability of AGTO in this regard. Figure 3e: Voltage Deviation Minimization Convergence Characteristics: This demonstrates how AG-TO converges in reducing voltage variations within the network. The graph shows that the algorithm is used to control voltage levels and thus works toward making the power system reliable and safe. Figure 3f: Voltage Stability Index (L-index) Minimization Convergence Characteristics—This shows how AGTO increases voltage stability through the reduction of the L-index. A lower L-index for a system is indicative of sound voltage stability and the graph shows that AGTO is efficient in maintaining stable voltage level within the system. Figure 3 illustrates the overall performance of the AGTO in terms of the mentioned objectives in the IEEE 30-bus test system. These Fig. 3a–f provide a general view that the algorithm is able to converge to the optimal or near-optimal solution to the OPF problem as well as address various challenges associated with the problem.

Using the AGTO in the formulation of the OPF problem provided favorable results. The AGTO algorithm thus effectively identified the best solutions that met the total generation cost minimizing objective as well as all the operational and security constraints. The solutions also proved to be quite resilient to contingency scenarios as well. The algorithm showed rather good convergence characteristics and provided nearly optimal solutions within affordable iterations. This was seen in the results of the AGTO as it was able to explore the solution space while at the same time being able to exploit the best solutions without getting stuck at local optima. The outcomes from this work validate the proposed AGTO approach in terms of providing accurate and efficient solutions for the real-world security-constrained AC OPF problem. It stresses the capabilities of AGTO to be a useful tool in power system planning and management. The above discussions lead to the conclusion that AGTO outperforms other algorithms due to the following features that are not found in other algorithms: Due to these characteristics, AGTO is well suited to solve complex and non-linear problems such as the OPF problem. However, like any other optimization algorithm, there are certain limitations to AGTO. As with any other algorithm, the efficiency of AGTO depends on the setting of certain parameters, including population size and number of iterations. However, this research shows that AGTO can be very beneficial to enhance the efficiency and security of power system management. It enhances the current research toward the development of effective and efficient optimization techniques for power systems and creates new horizons for the application of AGTO in other facets of power system optimization.

5 Conclusion

The Optimum Power Flow (OPF) problem can be effectively addressed utilizing the suggested AGTO method, as discussed in this paper. The AGTO method serves to solve OPF issues by striving to meet a range of active and reactive power goals. This algorithm's performance has been rigorously evaluated on the IEEE-30 Bus system, with the resultant data assessed against those obtained from other methods cited in relevant literature. The convergence speed of the AGTO is also examined through convergence curves provided across multiple case studies. When dealing with such challenging nonlinear problems, the convincing results point to AGTO providing superior solutions. Hence, AGTO emerges as a compelling alternative to conventional optimizers frequently deployed in a host of real-world OPF scenarios. This suggests that the OPF problem can indeed be tackled with the aid of the AGTO. Readers are encouraged to further enhance the efficacy of the AGTO approach in dealing with more complex, large-scale issues. This could be accomplished by amalgamating it with other algorithms or by integrating various enhancing strategies.

This research has implications for the power industry and can significantly influence the development of energy supply systems. It presents a powerful and effective method to solve power system optimization problems, which is essential for the economic and secure functioning of the power systems. By increasing the efficiency and security of power system control, AGTO could bring considerable cost benefits to power companies as well as improve the system reliability. In addition, the feature of AGTO in dealing with risks and opportunities will enable the integration of RES and DR, and help the transformation of power systems to more sustainable and reliable systems. Future work should extend on enhancing the performance of AGTO in the OPF problem. This may include designing of efficient parameter optimization strategies, incorporation of application dependent operators and improvement of management of uncertainties and exceptions. Furthermore, the use of AGTO in other power system optimization problems like unit commitment, power system planning, and renewable energy integration could be considered.

Also, future work can include the integration of AGTO with other optimization algorithms or machine learning methods. Such hybrid algorithms could potentially combine the advantages of each of the methods and yield even higher performance. In conclusion, this study offers fresh insights into the utilization of AGTO in power system optimization and lays a strong groundwork for further research in this field.

References

1. Carpentier JL (1985) Optimal power flows: uses, methods and developments. IFAC Proc Volumes 18(7):11–21. https://doi.org/10.1016/S1474-6670(17)60410-5
2. Niu M et al (2014) A review on applications of heuristic optimization algorithms for optimal power flow in modern power systems. J Mod Power Syst Clean Energy 2(4):289–297. https://doi.org/10.1007/s40565-014-0089-4
3. Kılıç U, Ayan K (2013) Optimizing power flow of AC-DC power systems using artificial bee colony algorithm. Int J Electr Power Energy Syst 53:592–602. https://doi.org/10.1016/j.ijepes.2013.05.036
4. Dutta S et al (2015) Hybrid biogeography-based optimisation for optimal power flow incorporating FACTS devices. Int J Power Energy Convers 6(1):63–84. https://doi.org/10.1504/IJPEC.2015.068423
5. Bouchekara HREH (2014) Optimal power flow using black-hole-based optimization approach. Appl Soft Comput 24:879–888. https://doi.org/10.1016/j.asoc.2014.08.056
6. Khaled U et al (2017) Optimal power flow using particle swarm optimization of renewable hybrid distributed generation. Energies 10(7):1013. https://doi.org/10.3390/en10071013
7. Sayah S, Hamouda A (2019) Optimal power flow solution of integrated AC-DC power system using enhanced differential evolution algorithm. Int Trans Electr Energy Syst 29(2):e2737. https://doi.org/10.1002/etep.2737
8. Bouchekara HRE-H, Abido MA (2014) Optimal power flow using differential search algorithm. Electr Power Compon Syst 42(15):1683–1699. https://doi.org/10.1080/15325008.2014.949912
9. Vaisakh K et al (2013) Genetic evolving ant direction particle swarm optimization algorithm for optimal power flow with non-smooth cost functions and statistical analysis. Appl Soft Comput 13(12):4579–4593. https://doi.org/10.1016/j.asoc.2013.07.002
10. Vaisakh K, Srinivas LR (2011) Genetic evolving ant direction HDE for OPF with non-smooth cost functions and statistical analysis. Expert Syst Appl 38(3):2046–2062. https://doi.org/10.1016/j.eswa.2010.07.142
11. Bakirtzis G et al (2002) Optimal power flow by enhanced genetic algorithm. IEEE Trans Power Syst 17(2):229–236. https://doi.org/10.1109/TPWRS.2002.1007886
12. Yuryevich J, Wong KP (1999) Evolutionary programming based optimal power flow algorithm. IEEE Trans Power Syst 14(4):1245–1250. https://doi.org/10.1109/59.801880
13. Mahdad B et al (2008) Optimal power flow of the Algerian network using genetic algorithm/fuzzy rules. In Proceedings of IEEE power and energy society general meeting-conversion and delivery of electrical energy in the 21st Century, place, pp 1–8. https://doi.org/10.1109/PES.2008.4596656
14. Leung HC, Chung TS (2000) Optimal power flow with a versatile FACTS controller by genetic algorithm approach. In: Proceedings of IEEE power engineering society winter meeting, place
15. Woo JH et al (2020) Real-time optimal power flow using twin delayed deep deterministic policy gradient algorithm. IEEE Access 8:213611–213618. https://doi.org/10.1109/ACCESS.2020.3041007
16. Duman S et al (2012) Optimal power flow using gravitational search algorithm. Energy Convers Manag 59:86–95. https://doi.org/10.1016/j.enconman.2012.02.024
17. Tan Y et al (2015) Improved group search optimization method for optimal power flow problem considering valve-point loading effects. Neurocomputing 148:229–239. https://doi.org/10.1016/j.neucom.2013.09.065
18. Shaheen et al (2022) Developed Gorilla troops technique for optimal power flow problem in electrical power systems. Mathematics 10(10):1636. https://doi.org/10.3390/math10101636
19. Hassan HA, Zellagui M (2018) Application of grey wolf optimizer algorithm for optimal power flow of two-terminal HVDC transmission system. Adv Electr Electron Eng 15(5):701–712. https://doi.org/10.15598/aeee.v15i5.2110

20. Shaheen M et al (2022) An improved heap optimization algorithm for efficient energy management based optimal power flow model. Energy 250:123795. https://doi.org/10.1016/j.energy.2022.123795
21. Sinsuphan N et al (2013) Optimal power flow solution using improved harmony search method. Appl Soft Comput 13(5):2364–2374. https://doi.org/10.1016/j.asoc.2013.01.024
22. Ben Hmida JB et al (2019) Solving constrained optimal power flow with renewables using hybrid modified imperialist competitive algorithm and sequential quadratic programming. Electr Power Syst Res 177:105989. https://doi.org/10.1016/j.epsr.2019.105989
23. Kahraman HT et al (2022) Optimization of optimal power flow problem using multi-objective manta ray foraging optimizer. Appl Soft Comput 116:108334. https://doi.org/10.1016/j.asoc.2021.108334
24. Mohamed A-AA et al (2017) Optimal power flow using moth swarm algorithm. Electr Power Syst Res 142:190–206. https://doi.org/10.1016/j.epsr.2016.09.025
25. Niknam T et al (2011) A modified shuffle frog leaping algorithm for multi-objective optimal power flow. Energy 36(11):6420–6432. https://doi.org/10.1016/j.energy.2011.09.027
26. Bentouati B et al (2016) A solution to the optimal power flow using multi-verse optimizer. J. Electr. Syst. 12:716–733
27. Weng X et al (2023) A vertical and horizontal crossover sine cosine algorithm with pattern search for optimal power flow in power systems. Energy 271:127000. https://doi.org/10.1016/j.energy.2023.127000
28. Rahiminejad et al (2014) Shuffled frog leaping algorithm optimization for AC–DC optimal power flow dispatch. Turk J Elec Eng Comp Sci 22:874–892. https://doi.org/10.3906/elk-1205-80
29. Farhat M et al (2022) ESMA-OPF: enhanced slime mould algorithm for solving optimal power flow problem. Sustainability 14(4):2305. https://doi.org/10.3390/su14042305
30. Alanazi et al (2022) Determining optimal power flow solutions using new adaptive Gaussian TLBO method. Appl Sci 12(16):7959. https://doi.org/10.3390/app12167959
31. Ongsakul W, Bhasaputra P (2002) Optimal power flow with FACTS devices by hybrid TS/SA approach. Int J Electr Power Energy Syst 24(10):851–857. https://doi.org/10.1016/S0142-0615(02)00006-6
32. Logeswaran T et al (2022) Power flow management of hybrid system in smart grid requirements using ITSA-MOAT approach. Appl Energy 319:119228. https://doi.org/10.1016/j.apenergy.2022.119228
33. Roy R, Jadhav HT (2015) Optimal power flow solution of power system incorporating stochastic wind power using Gbest guided artificial bee colony algorithm. Int J Electr Power Energy Syst 64:562–578. https://doi.org/10.1016/j.ijepes.2014.07.010
34. Panda A, Tripathy M (2014) Optimal power flow solution of wind integrated power system using modified bacteria foraging algorithm. Int J Electr Power Energy Syst 54:306–314. https://doi.org/10.1016/j.ijepes.2013.07.018
35. Panda A, Tripathy M (2015) Security constrained optimal power flow solution of wind-thermal generation system using modified bacteria foraging algorithm. Energy 93:816–827. https://doi.org/10.1016/j.energy.2015.09.083
36. Shi L et al (2012) Optimal power flow solution incorporating wind power. IEEE Syst J 6(2):233–241. https://doi.org/10.1109/JSYST.2011.2162896
37. Mishra S et al (2011) Security constrained economic dispatch considering wind energy conversion systems. In: Proceedings of IEEE power and energy society general meeting, place, pp 1–8. https://doi.org/10.1109/PES.2011.6039544
38. Zhou W et al (2011) Optimal wind-thermal coordination dispatch based on risk reserve constraints. Eur Trans Electr Power 21(1):740–756. https://doi.org/10.1002/etep.474
39. Dubey HM et al (2015) Hybrid flower pollination algorithm with time-varying fuzzy selection mechanism for wind integrated multi-objective dynamic economic dispatch. Renew Energy 83:188–202. https://doi.org/10.1016/j.renene.2015.04.034
40. Tazvinga H et al (2015) Optimal power flow management for distributed energy resources with batteries. Energy Conversat Manag 102:104–110

41. Kusakana K (2016) Optimal scheduling for distributed hybrid system with pumped hydro storage. Energy Conversat Manag 111:253–260
42. Reddy Salkuti S et al (2018) Multi-objective based optimal generation scheduling considering wind and solar energy systems. Int J Emerg Electr Power Syst 19(5):20180006. https://doi.org/10.1515/ijeeps-2018-0006
43. Reddy SS, Momoh JA (2015) Realistic and transparent optimum scheduling strategy for hybrid power system. IEEE Trans Smart Grid 6(6):3114–3125. https://doi.org/10.1109/TSG.2015.2406879
44. Surender Reddy S et al (2015) Real-time economic dispatch considering renewable power generation variability and uncertainty over scheduling period. IEEE Syst J 9(4):1440–1451. https://doi.org/10.1109/JSYST.2014.2325967
45. Salkuti SR (2019) Optimal operation of microgrid considering renewable energy sources, electric vehicles and demand response. In: E3S web conference, vol 87, p 01007
46. Pandya SB, Jariwala HR (2021) A different perception of hybrid renewable energy sources integrated multi-objective optimal power flow considering performance parameters and penetration. Smart Sci. 9(3):186–215. https://doi.org/10.1080/23080477.2021.1916853
47. Pandya SB, Jariwala HR (2021) Equilibrium optimizer: Insights, balance, diversity for renewable energy resources based optimal power flow with multiple scenarios. Smart Sci 9(4):257–274. https://doi.org/10.1080/23080477.2021.1932164
48. Biswas PP et al (2018) Multiobjective economic-environmental power dispatch with stochastic wind-solar-small hydro power. Energy 150:1039–1057. https://doi.org/10.1016/j.energy.2018.03.002
49. Agrawal S, Kundu P (2021) A novel multi-objective unified optimal power flow based methodology for optimal installation of VSC-HVDC converter. Int Trans Electr Energy Syst 31(8):e12956. https://doi.org/10.1002/2050-7038.12956
50. Abdollahzadeh B, Soleimanian Gharehchopogh F, Mirjalili S (2021) Artificial gorilla troops optimizer: a new nature-inspired metaheuristic algorithm for global optimization problems. Int J Intell Syst 36(10):5887–5958. https://doi.org/10.1002/2050-7038.12956

Boosting Accuracy and Transparency in Biometric Finger-Vein Recognition: CNN with Adaptive Interpretable Recognition

Manvi Khatri and Ajay Sharma

Abstract Biometric finger-vein recognition is a promising approach for secure and reliable identification of individuals based on the unique patterns present in the veins of their fingers. However, achieving high accuracy, adaptability, and interpretability in finger-vein recognition systems remains a challenge. In this study, we propose a CNN (Convolutional Neural Network) with adaptive interpretable recognition along transfer learning approach to address these challenges. First, we leverage the power of CNNs by employing the Xception model, a state-of-the-art deep learning architecture, as the backbone of our system. The Xception model is pre-trained on large-scale datasets and provides excellent feature extraction capabilities, capturing discriminative finger-vein patterns effectively. To enhance adaptability, we employ transfer learning techniques. By starting with a pre-trained Xception model, we leverage the learned feature representations and fine-tune the model using a smaller labeled dataset specific to finger-vein recognition. This approach reduces the need for a large amount of labeled data and improves the model's generalization ability. In addition to accuracy, interpretability plays a crucial role in biometric systems. To address this, we integrate interpretable recognition techniques. By combining the power of CNNs, adaptive transfer learning, and interpretable recognition, we achieve high accuracy, adaptability to varying finger-vein patterns, and interpretability for user trust and understanding. This work contributes to the development of robust and transparent biometric recognition systems, with potential applications in security, access control, and identity verification.

Keywords CNN (Convolutional Neural Network) · Adaptive recognition · Transfer learning · Biometric recognition · Finger-vein recognition · Xception model · Deep learning · Attention mechanisms · PolyU Finger Vein Dataset · Accuracy

---

M. Khatri · A. Sharma (✉)
Department of Computer Science and Engineering, SRM University, Delhi-NCR, Sonepat, India
e-mail: ajay.s@srmuniversity.ac.in

A. Kumar et al. (eds.), *Trends in Mechatronics Systems*, Emerging Trends in Mechatronics, https://doi.org/10.1007/978-981-97-9108-8_3

1 Introduction

Biometrics is the scientific study of identifying and verifying persons via the use of their measurable biological or behavioral traits. The emphasis is on establishing a person's identity through their unique characteristics. Because of the difficulty in forging or recreating such features, biometric systems rely on them. Both physiological and behavioral characteristics are considered biometric features [1, 2].

- **Physiological Biometrics**: Physical qualities of the body serve as the basis for these characteristics. Fingerprints, irises, faces, hands, veins, retinas, and other physiological characteristics are all examples of biometrics. To verify a person's identification, physiological biometrics analyze their bodily characteristics.
- **Behavioral Biometrics**: These traits are based on the unique behavioral patterns exhibited by individuals. Examples of behavioral biometrics include voice patterns, signature dynamics, keystroke dynamics, gait analysis, and mouse movement. Behavioral biometrics focus on how individuals interact with their environment and exhibit distinct patterns in their actions.

Biometric systems capture and measure these traits using specialized sensors or devices, such as fingerprint scanners, iris scanners, cameras, microphones, or other relevant sensors. The captured biometric data is converted into digital templates or mathematical representations that can be stored, compared, and matched against existing records in a database [3–5].

The primary goal of biometrics is to establish or confirm the identity of an individual accurately and reliably. Biometric technology finds applications in various sectors, including law enforcement, border control, access control to secure facilities, identity verification for financial transactions, attendance management, and personal device authentication. It is important to note that the implementation of biometric systems must address privacy concerns and comply with data protection regulations. Safeguards such as secure storage and transmission of biometric data, informed consent, and appropriate handling of sensitive information are essential to maintain privacy and ensure ethical use of biometric technology. Biometric traits are the unique physical or behavioral characteristics of individuals that can be measured, analyzed, and used for biometric identification, verification, or authentication. Here are some commonly used biometric traits:

Physiological Biometric Traits:

- *Fingerprint*: Analysis of the patterns formed by ridges and valleys on the fingertips.
- *Iris*: Examination of the unique patterns in the colored portion of the eye surrounding the pupil.
- *Face*: Capture and analysis of facial features, including the distance between the eyes, nose shape, and jawline.
- *Palmprint*: Focus on the unique patterns on the palm of the hand.
- *Retina*: Analysis of the patterns of blood vessels at the back of the eye.

- *Hand Geometry*: Measurement of the size and shape of the hand and fingers.
- *Vein Pattern*: Analysis of the patterns of veins in the hand, palm, or other body parts.
- *Ear Shape*: Analysis of the unique shape and structure of the ear [6, 7].

Behavioral Biometric Traits:

- *Voice*: Analysis of the unique characteristics of an individual's voice, including pitch, tone, and speech patterns.
- *Signature*: Examination of an individual's handwriting style and stroke patterns.
- *Keystroke Dynamics*: Measurement of an individual's typing rhythm, including keystroke timing and pressure.
- *Mouse Movement*: Analysis of the unique patterns in an individual's mouse usage, including speed and acceleration.
- *Gait*: Analysis of an individual's walking pattern, including stride length, arm swing, and posture.
- *Gesture*: Capture and analysis of specific hand or body movements for identification or authentication purposes.

It is important to note that these are examples of commonly used biometric traits, but the field of biometrics is dynamic, and new traits or modalities may emerge as research and development progress [8]. Additionally, different applications or systems may prioritize specific biometric traits based on factors such as accuracy, convenience, or suitability for the intended use case. Biometric technology has witnessed significant advancements, but it still faces several challenges that need to be addressed to ensure its effectiveness, reliability, and ethical use. Here are some common challenges in the field of biometrics:

- *Accuracy and Reliability*: Biometric systems must achieve high accuracy and reliability in identifying or verifying individuals. Factors such as variations in biometric traits, changes over time, and environmental conditions can impact the performance of biometric systems. Ensuring consistent and accurate results is a continual challenge.
- *Data Quality*: Biometric data quality is crucial for reliable identification. Factors like poor image resolution, sensor limitations, or improper capture techniques can lead to degraded data quality, resulting in reduced system performance. Maintaining high-quality data is essential for accurate biometric analysis.
- *Privacy and Security*: Biometric data is highly sensitive as it uniquely identifies individuals. Safeguarding biometric data, ensuring secure storage and transmission, and implementing appropriate access controls are critical to protecting privacy and preventing unauthorized access or misuse.
- *Template Protection*: Biometric systems typically store templates or mathematical representations derived from biometric data instead of storing the original data itself. Protecting these templates from unauthorized access or reverse engineering is essential to maintain the security and integrity of the system.
- *Ethical Considerations*: Biometrics raises ethical concerns related to consent, transparency, and potential biases. Ensuring fairness, non-discrimination, and

transparency in the collection, storage, and use of biometric data is crucial to maintain public trust and uphold ethical standards.

- *Interoperability and Standardization*: Biometric systems from different vendors or organizations may have interoperability issues, hindering the exchange or integration of biometric data. Establishing common standards and protocols is necessary to ensure compatibility and seamless interoperability between different biometric systems.
- *User Acceptance and Usability*: User acceptance plays a vital role in the success of biometric systems. Factors such as user perception of security and privacy, convenience, ease of use, and cultural considerations can influence user acceptance. Designing user-friendly interfaces and addressing user concerns are essential to encourage adoption and user satisfaction.
- *Scalability*: Biometric systems must handle large-scale databases with an increasing number of enrolled individuals. Ensuring efficient search and matching operations, managing storage requirements, and maintaining system performance as the database grows are significant scalability challenges [9–11].

Addressing these challenges among researchers, industry experts, and policy-makers are main objectives of this research. There are following research questions for this research as below:

1. How can deep learning techniques, particularly CNNs, be effectively applied to biometric finger-vein recognition to improve accuracy and reliability?
2. What are the key challenges in finger-vein recognition, and how can adaptive recognition objectives be designed to address these challenges?
3. What is the impact of dataset size and diversity on the accuracy of CNN-based finger-vein recognition systems?
4. How can interpretable neural network architectures be designed for finger-vein recognition to enhance transparency in decision-making?

Overcoming these obstacles will contribute to the advancement and responsible deployment of biometric technology in various applications. Researchers have contributed significantly to the area by studying how to combine Convolutional Neural Networks (CNNs) with adaptive interpretable recognition objectives to improve the accuracy and transparency of biometric finger-vein detection. Key contributions to knowledge that may be gained from such a study include the following:

- To improve the reliability and security of finger-vein recognition, this research can help by creating and assessing new recognition objectives and CNN architectures.
- Transparency in biometric recognition may be improved by developing and exploring approaches to depict and explain model decisions, which in turn increases user confidence and comprehension.
- Finger-vein recognition systems that are flexible enough to perform well on a wide variety of datasets can be made possible by research that addresses the difficulty of cross-database recognition.

- The study can investigate methods to make finger-vein identification models more resilient to variations in illumination, finger orientation, and picture quality, making the system more trustworthy in real-world settings.
- The security and trustworthiness of the technology is improved by studies that aim to make finger-vein recognition systems more resistant to adversarial assaults, especially in contexts where security is of the utmost importance.
- Developing standards and best practices for the ethical use of biometric data in recognition systems and addressing related ethical and privacy concerns are important steps toward assuring the responsible and transparent deployment of technology.

Improving system performance, security, and ethical considerations through research into enhancing accuracy and transparency in biometric finger-vein recognition with CNNs and interpretable recognition objectives can significantly advance the field and allow for the responsible and effective deployment of biometric recognition technology across a wide range of applications.

2 Review of Literature

In their review of the field of biometrics, Jain et al. [2] identify some of the most pressing research questions that must be answered before biometric technology can be used effectively to ensure the safety of sensitive data. In today's highly interdependent world, it's more important than ever to put a face to a name.

An experimental examination of the use of soft biometric labels based on the description of human physical attributes to enhance difficult distant person recognition scenarios is presented by Tome et al. [12].

Using just rudimentary sensor-side information of biometric faking, Menotti et al. [13] build two deep learning-based spoofing detection algorithms that perform very well across iris, face, and fingerprint modalities. Individual, national, and international safety have all benefited from biometrics systems' enhanced capacity for identifying and authenticating individuals.

A biometric cryptosystem, in which a cryptographic secret is hidden using biometric modalities, is presented in a novel framework by Kumar et al. (2015). In this study, we employ two methods to ensure the integrity of the candidate biometric modality: (1) BCH encoding, which provides the parity-code saved for the alignment of the query biometric template, and (2) the Hash function to calculate a hash-code.

In this research, Ali et al. [14] present a multimodal authentication system that makes use of encrypted biometrics in an edge-centric cloud setting. The suggested method employs the usage of mobile personal devices for biometric encryption, therefore maximizing resource utilization and addressing a further constraint of the cloud. A novel method of biometric encryption has been developed.

Khodadoust et al. [15] introduce a palm print and fingerprint based multibiometric system. In order to effectively integrate the discriminative strength of several biometric features, the suggested method is based on a score fusion. We suggest a hybrid approach to fingerprint image improvement, and the results are promising.

Wu et al. [16] show that the usage of a single biometric approach for wireless multimedia authentication is widespread. Spoofing attacks are possible, and its precision is low. To address this problem, we offer a dynamic Bayesian strategy for multimodal fusion of fingerprint.

Despite the advances made in face detection and identification over the past decade, Mohsin et al. [17] noted that study of facial features remains a hot research area. Putting aside full-face recognition, it is still seen to be productive to investigate the potentials of soft biometric qualities, such as individual facial attributes like the nose, the lips, the hair, and so on.

A constantly developing technology known as biometrics is currently being applied in a plethora of personal, government, and commercial applications to combat challenges including identity theft and security risks [18]. Despite significant advancements, some pressing issues must be resolved before biometrics can reach its full potential.

To improve accuracy and performance, Tiong et al. [19] show that the proposed network makes use of multifeature fusion layers and provides a discriminative feature representation. Surveillance applications that rely on facial recognition technology continue to face difficulties in uncontrolled contexts due to factors such as the prevalence of masks and veils and cultural differences in look. One of the primary areas of research to deal with such situations is multimodal face biometrics identification.

For the feature level fusion of fingerprint and online signature, Leghari et al. [20] propose and construct a Convolutional Neural Network (CNN) based model. The advent of contemporary technology and the substantial study of multimodal biometrics by the academic community have necessitated the use of multimodal biometrics in practical settings. Noise, reduced universality, intra-class variability, and spoof assaults are only some of the challenges faced by single-modality biometric systems.

To create a safe multimodal template using a user's face and iris biometrics, Talreja et al. [21] provide a deep learning system for feature-level fusion. To produce a reliable binary multimodal shared latent representation, we incorporate a deep hashing (binarization) approach into the fusion architecture.

Zhong et al. [22] investigated a palmprint-and-dorsal-hand-veins (DHV) multi-biometric approach. The biopsy (Liveness detection) is the main benefit of DHV recognition, and palmprint recognition is quite accurate and reliable. The following are some strengths and weaknesses of existing methods:

1. There is a dearth of in-depth studies examining how well finger-vein identification systems operate outside of controlled laboratory conditions. To evaluate the effectiveness of these technologies in real-world settings, more study is required.
2. Inadequate variety in existing finger-vein identification datasets may result in inaccurate models. More study is required to compile datasets that adequately represent society at large with regards to age, gender, and ethnicity.

3. Adversary assaults on biometric systems, such as finger-vein recognition, are a developing problem. There is a lack of knowledge about how vulnerable these systems are to assaults and how to protect them.
4. Ethical concerns around finger-vein recognition data collection, storage, and use have received little attention in the research literature. It is crucial to look into privacy-protecting methods that are both legal and ethical.
5. While research on how to most efficiently incorporate interpretability into biometric systems exists, there is still a need for it. It is critical to establish consistent methods for justifying model choices.
6. The generalizability of studies to other finger-vein datasets is typically overlooked in favor of investigating the dataset in question. Recognizing the constraints of cross-database identification algorithms is an important unexplored area of study.

Biometric finger-vein recognition systems may be made more reliable, secure, and widely applicable if the aforementioned knowledge gaps are identified and filled in.

3 Problem Formulation

Biometric finger-vein recognition is a form of biometrics that relies on an individual's fingerprints by analyzing the vein patterns in their fingers. It's a popular biometric method since it doesn't need much effort from the user and provides a high level of security. Some benefits of biometric finger-vein recognition are as follows. Each person has a distinctive pattern of veins in their fingers, which may be used to positively identify them. Images of finger veins may be taken without touching the person or causing any discomfort.

Secure: Vein patterns in the fingers are internal to the body and hence tough to fake. The fingerprint vein pattern is not very sensitive to changes in temperature and pressure. Finger vein patterns are less affected by aging than other biometric traits such as fingerprints.

Some drawbacks of biometric finger-vein recognition include the following: Finger-vein scanners still have a hefty price tag. Vein scanners for the fingers are not as common as fingerprint scanners, yet both are biometric methods. It is possible for finger-vein patterns to be concealed by things like dirt and filth. Changes in the finger-vein pattern can occur with age, injury, or illness. There are a number of benefits that make biometric finger-vein recognition a viable biometric technique [18, 23–26]. However, it is important to be aware of the limitations of the technology before deploying it in a security application. The problem formulation of biometric finger-vein recognition can be summarized as follows:

Input:

A finger-vein image of an individual.

Output:

A decision of whether the individual is the legitimate user or an impostor.

Steps:

- *Image acquisition*: The finger vein image is acquired using a finger vein scanner.
- *Image preprocessing*: The image is preprocessed to remove noise and artifacts.
- *Feature extraction*: Features are extracted from the image, such as the shape, size, and texture of the veins.
- *Matching*: The features are matched against a database of known finger vein images to determine whether the individual is a legitimate user or an impostor.

The success of biometric finger-vein recognition depends on the following factors:

- Image quality.
- Feature extraction method.
- Matching algorithm.

Biometric finger-vein recognition is a promising biometric technology with a number of advantages over other biometric technologies, such as:

- *High accuracy*: Finger-vein patterns are unique to each individual and can be used to identify individuals with high accuracy.
- *Non-intrusive*: The process of capturing finger-vein images is non-intrusive and does not require physical contact with the individual.
- *Secure*: Finger-vein patterns are located inside the body and are difficult to forge.

However, biometric finger-vein recognition also has some limitations, such as:

- *Cost*: The cost of finger-vein scanners is still relatively high.
- *Environmental factors*: Finger-vein patterns can be affected by environmental factors such as temperature and pressure.
- *Variations*: Finger-vein patterns can vary over time due to factors such as aging and injury.

Overall, biometric finger-vein recognition is a promising biometric technology with a number of advantages over other biometric technologies. However, it is important to be aware of the limitations of the technology before deploying it in a security application [27]

4 Dataset

The PolyU Finger-Vein Dataset, developed by The Hong Kong Polytechnic University, is a widely recognized and commonly used dataset for research and evaluation of finger-vein recognition systems. It provides a valuable resource for developing and testing finger-vein recognition algorithms. The PolyU Finger-Vein Dataset consists of finger-vein images captured using a near-infrared imaging system. The dataset contains multiple samples of finger-vein images from hundreds of individuals [15, 28–30].

The dataset includes a substantial number of finger-vein samples, providing ample data for training, validation, and testing purposes. The exact number of individuals and samples may vary depending on the specific version or subset of the dataset being used. The dataset covers a range of finger positioning variations, including different finger placements and angles. It also includes variations in lighting conditions, providing challenges similar to real-world scenarios. The PolyU Finger Vein Dataset typically comes with pre-defined training and testing subsets, allowing researchers to evaluate the performance of finger-vein recognition algorithms on standardized evaluation protocols. These subsets ensure fair and consistent evaluation across different approaches [31–35].

5 Work Done

Integrating CNN with adaptive interpretable recognition and transfer learning for biometric finger-vein recognition involves combining these approaches into a cohesive system. Then fine-tuned or adapted for a different target task with limited labeled data. These pre-trained models serve as a starting point and provide learned representations that can be utilized to improve the performance of the target task. Adaptive interpretable recognition refers to a biometric recognition system that is capable of dynamically adapting to changing conditions and providing interpretable insights into its decision-making process. It combines the principles of adaptability and interpretability to enhance the performance, robustness, and transparency of the recognition system.

Here's a step-by-step guide on how you can approach this integration:

Step 1. *Data Preprocessing:* This step ensure the dataset is properly labeled with corresponding identity or class labels.

Step 2. *Transfer Learning with CNN:* This step begin by utilizing transfer learning with a pre-trained CNN model, such as a widely used architecture like ResNet or VGGNet. Import the pre-trained model and remove the last fully connected layer.

Step 3. *Fine-tuning:* This step unfreeze a subset of the pre-trained layers or the entire network, allowing them to be fine-tuned using the finger-vein dataset. Train the model using the labeled finger-vein images, adjusting the model's weights to adapt to the specific vein patterns present in the dataset.

Step 4. *Adaptive Recognition:* This step implement adaptive recognition techniques to account for intra-individual variations in finger-vein patterns. This can involve periodically updating the recognition model using new finger-vein images from enrolled users. Incorporate techniques such as online or incremental learning to adapt the model to changes over time, improving its accuracy and robustness.

Step 5. *Interpretable Recognition:* This step utilize interpretable recognition techniques to gain insights into the decision-making process of the CNN model. This can involve visualizing and analyzing the activations or attention maps of the model to identify the vein patterns or regions that contribute most to the recognition decision. Implement methods such as Grad-CAM or saliency maps to visualize the important regions within the finger-vein images.

Step 6. *Evaluation and Performance Metrics:* This step ensure to assess the system's accuracy, robustness, and adaptability to intra-individual variations.

Step 7. *Model Optimization and Refinement:* This step fine-tune the integrated model based on evaluation results and analysis. Adjust hyperparameters, explore different network architectures, or incorporate additional techniques to further optimize the performance of the system.

Step 8. *Validation and Testing:* This step validates the integrated system on a separate validation set to ensure it performs consistently. Finally, assess the system's performance on the testing set to obtain a reliable evaluation of its accuracy and robustness.

Regular updates and maintenance of the system are also important to ensure continuous adaptability and performance improvement. The combination of adaptability and interpretability in biometric recognition systems enables the system to learn from new data, handle variations, and provide transparency in the decision-making process. These capabilities enhance the system's performance, robustness, and trustworthiness, making it suitable for real-world applications where biometric data may change over time or exhibit variations [36].

6 Results and Analysis

In Keras, the authors load the Xception model using the Xception function. This function takes a number of arguments, including whether to include the top layer of the network (which contains the classification layer), the weights of the network (either imagenet for the pre-trained weights or a path to a custom weights file), and the input shape of the images. Once it has been loaded the model, the authors pass an image to the proposed model's predict method to get the predicted class label for the image.

Some common preprocessing techniques for finger vein images include noise removal: This can be done using a variety of methods, such as Gaussian blurring, median filtering, or wavelet denoising. Vein enhancement can be done using a variety of methods, such as morphological operations, thresholding, or edge detection. The specific preprocessing techniques that are used will depend on the specific characteristics of the finger vein images and the desired results in Fig. 1.

Here are some of the benefits of finger vein preprocessing:

- It can make finger vein recognition systems more robust to noise and artifacts.

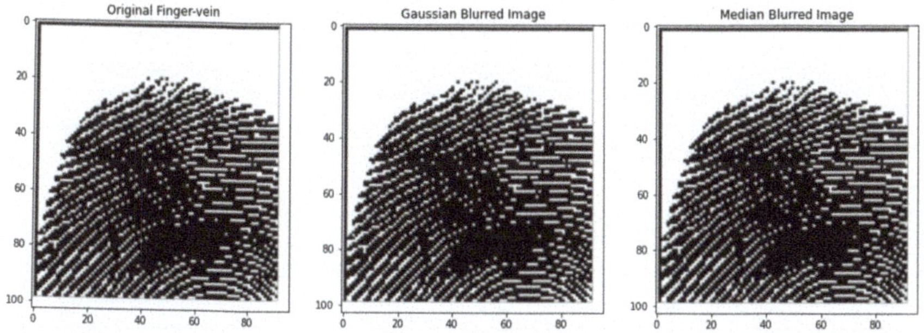

Fig. 1 Preprocessing

- Here are some of the challenges of finger vein preprocessing:

The quality of finger vein images can vary significantly, depending on factors such as the skin condition of the user, the lighting conditions, and the position of the finger in Fig. 2.

Despite the challenges, finger vein preprocessing is an important step in the finger vein recognition process. By removing noise and artifacts, and enhancing the visibility of the vein patterns, preprocessing can improve the accuracy, robustness, and usability of finger vein recognition systems. Visualizing the loss curve can help in understanding the training progress, identifying potential issues like overfitting, and determining the optimal stopping point for training. It is typically plotted with the number of training iterations or epochs on the x-axis and the loss values on the y-axis. The specific shape of the loss curve can vary depending on factors such as the dataset, model architecture, training parameters, and regularization techniques employed in Fig. 3.

It's essential to monitor the loss curve and make appropriate adjustments during the training process to achieve the desired performance in biometric finger-vein recognition in Fig. 4.

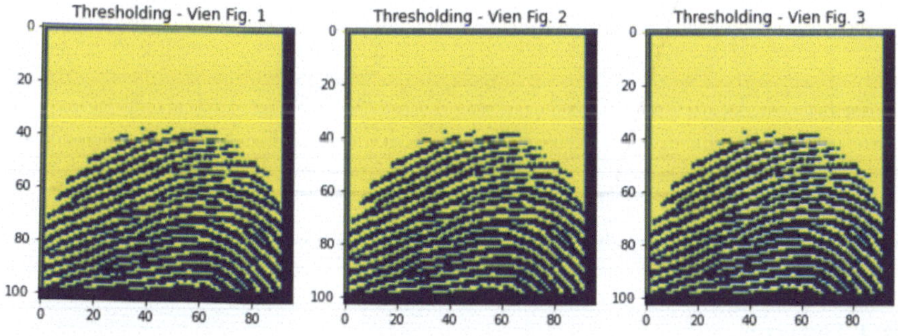

Fig. 2 Threshold

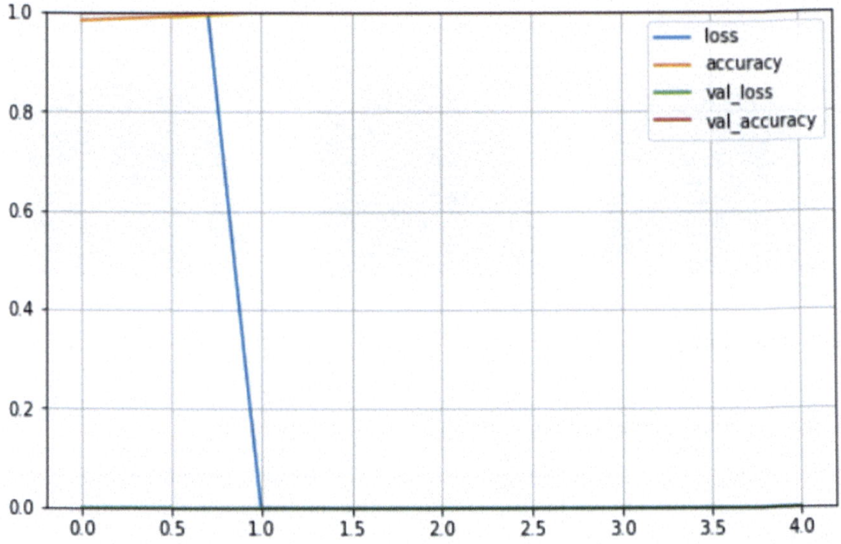

Fig. 3 Performance evaluation

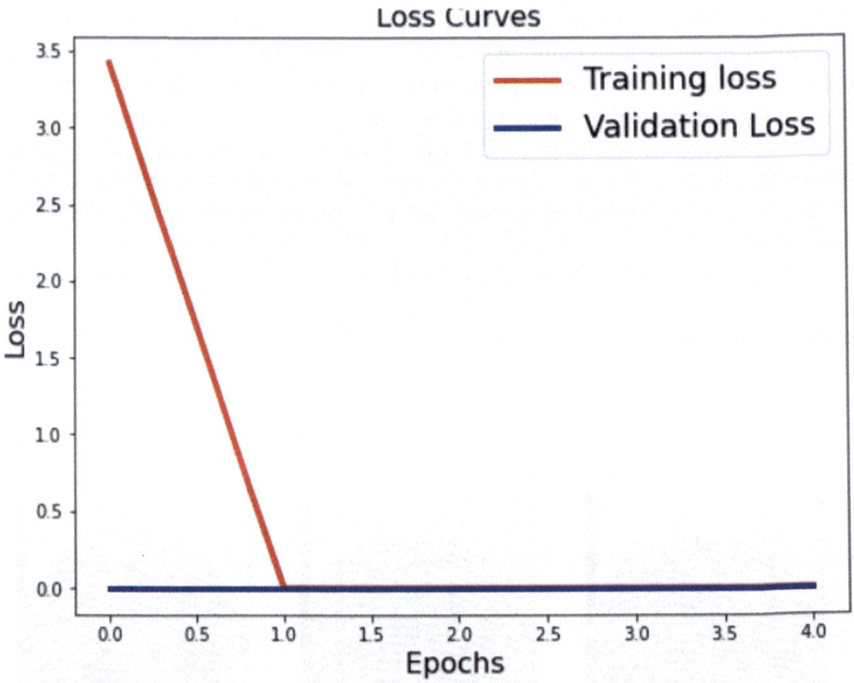

Fig. 4 Loss curve

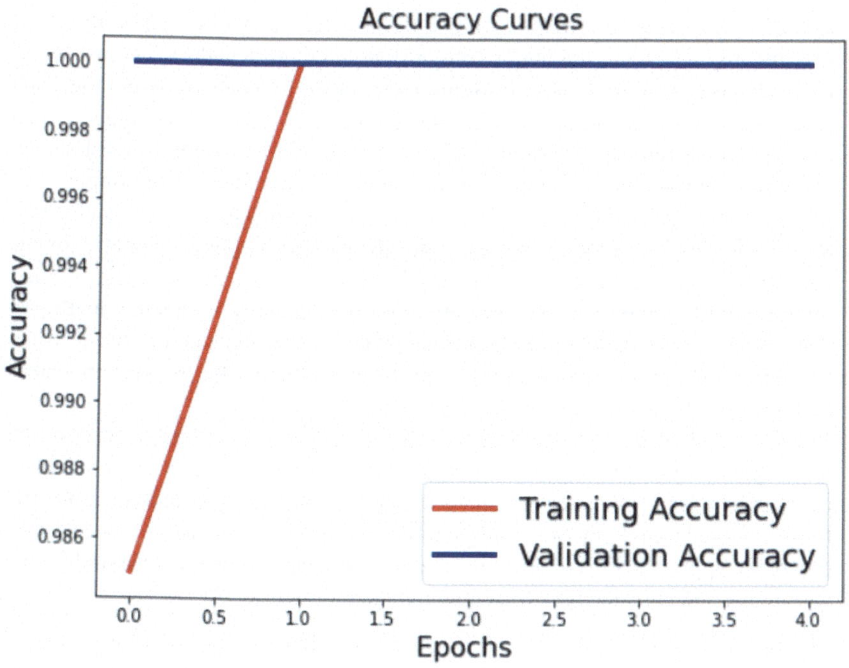

Fig. 5 Accuracy curve

The loss curve in biometric finger-vein recognition represents the training and validation loss values over the course of the model's training process. It provides insights into how the model's performance improves or converges during training in Fig. 5.

Visualizing the accuracy curve can help in understanding the training progress, identifying potential issues like overfitting, and determining the optimal stopping point for training. It is typically plotted with the number of training iterations or epochs on the x-axis and the accuracy values on the y-axis [37–41]. CNN with adaptive interpretable recognition along transfer learning for biometric finger-vein recognition is an approach that combines several techniques to improve the performance, adaptability, and interpretability of the recognition system. Let's discuss the key aspects of this approach:

- *CNN (Convolutional Neural Network)*: By utilizing CNNs for biometric finger-vein recognition, we can benefit from their ability to capture discriminative patterns and representations.
- *Transfer Learning*: By starting with a pre-trained CNN model, we can benefit from the learned feature representations, which often capture general visual patterns. We then adapt or fine-tune the pre-trained model to the specific task of biometric finger-vein recognition using a smaller labeled dataset.

- *Adaptive Recognition*: Adaptive recognition refers to the system's ability to dynamically adapt and update its recognition model as new data becomes available. This is crucial for handling variations in finger-vein patterns over time, such as changes due to aging or different finger positions. Adaptive techniques, such as online or incremental learning, enable the recognition system to continuously learn and improve its performance as it encounters new finger-vein samples.
- *Interpretable Recognition*: Interpretable recognition aims to provide insights into the decision-making process of the recognition system. This is particularly important for building trust and understanding in biometric systems. By incorporating interpretability techniques, such as attention mechanisms or saliency maps, we can visualize and analyze the important regions or features contributing to the recognition decision. This enables users to understand why the system makes certain predictions and enhances transparency.
- *Integration of Techniques*: By integrating adaptive and interpretable recognition techniques along with transfer learning, we create a system that can adapt to changing finger-vein patterns, utilize learned feature representations from pre-trained models, and provide interpretable explanations for its decisions. This combined approach aims to improve recognition accuracy, handle intra-individual variations, and enhance user trust and understanding [42–44].

The discussion of CNN with adaptive interpretable recognition along transfer learning for biometric finger-vein recognition highlights the importance of leveraging deep learning models, incorporating adaptive mechanisms, and providing interpretable explanations. There are several real-world applications for improving the accuracy and openness of biometric finger-vein recognition utilizing Convolutional Neural Networks (CNNs) with adaptive interpretable recognition. Some real-world examples and ramifications are as follows:

- Organizations, data centers, and other high-security facilities can include biometric finger-vein recognition into their secure access control systems to increase security and reduce dependency on manual entry methods like cards and PINs.
- Finger-vein recognition can be utilized for identity verification in e-commerce and online services, making them safer by lowering the likelihood of fraud and identity theft.
- Financial organizations like banks may protect their customers' personal and financial information via finger-vein recognition authentication.
- Finger-vein recognition can be used to verify the identity of medical staff, grant them access to patient records, and protect private health information.
- Finger-vein recognition can be used by law enforcement authorities for the purpose of identifying criminals and verifying the identities of suspects.

Ethical, privacy, and operational concerns must be taken into account throughout the safe and efficient deployment of biometric finger-vein recognition systems in a wide range of practical settings. With careful planning and implementation, these technologies have the potential to vastly enhance safety and productivity across a variety of sectors.

By combining these techniques, we can develop more robust and transparent biometric recognition systems that adapt to changing conditions and provide insights into their decision-making process [45].

7 Conclusion and Future Scope

The integration of CNN with adaptive interpretable recognition along transfer learning for biometric finger-vein recognition offers a powerful approach to improve the performance, adaptability, and interpretability of the recognition system. By leveraging CNNs, we can effectively capture discriminative patterns from finger-vein images and utilize pre-trained models to benefit from learned feature representations. Transfer learning allows us to transfer knowledge from large-scale datasets to the target task of finger-vein recognition, reducing the need for a large labeled dataset. This enables us to fine-tune the pre-trained model and adapt it to the specific finger-vein recognition task, improving accuracy and generalization. The adaptive recognition component addresses the challenges of handling variations in finger-vein patterns over time. By incorporating online or incremental learning techniques, the system can continuously update its recognition model with new data, adapting to changes in finger-vein patterns, and improving performance. Interpretable recognition enhances transparency and user trust by providing insights into the system's decision-making process. By utilizing attention mechanisms, saliency maps, or other interpretability techniques, the system can visualize and explain the important regions or features contributing to the recognition decision. This allows users to understand and validate the system's predictions.

In order to improve accuracy and transparency in biometric finger-vein recognition, using Convolutional Neural Networks (CNNs) with adaptive interpretable recognition objectives is a promising technique. However, this method is not without its limits and obstacles.

- Model performance might be hampered by insufficient or poor quality data sources. One of the biggest obstacles is getting enough varied, high-quality finger-vein data.
- It's possible that models trained on a certain dataset won't transfer well to others or to the actual world. It is still difficult to guarantee generality across databases and settings.
- Even in an interpretable system, there are legitimate privacy concerns with the usage of biometric data. There will always be difficulties in securing biometric information and meeting privacy standards.
- It's difficult to create models that can be easily understood in the context of CNNs. There is need for improvement in obtaining full transparency, despite the fact that attention mechanisms and explainable AI strategies can assist.

- Variables such as finger alignment, illumination, and picture quality can all have an impact on the accuracy of finger-vein identification. There is a lack of robustness under these extreme situations.
- Finger-vein identification systems that rely on CNNs may be susceptible to assaults from malicious actors. The ability to withstand assaults and negative examples is a major problem.
- Users may be skeptical of or even hostile to biometric technologies like finger-vein recognition. The inability to identify and resolve user issues and preferences.
- When working with limited hardware or software, it might be difficult to train and deploy deep learning models. It might be difficult to find solutions that don't break the bank.

Research and development initiatives, cross-disciplinary activities, and a dedication to resolving ethical and privacy concerns are all necessary to advance biometric finger-vein recognition beyond its current limits. Despite their promising promise, implementations of CNNs with adaptive interpretable recognition targets should be done only after careful consideration of these constraints.

The integration of these techniques creates a comprehensive framework that not only achieves high accuracy in biometric finger-vein recognition but also ensures adaptability to changing conditions and provides interpretable explanations. This approach is crucial in building robust and transparent biometric systems, enhancing user trust, and enabling a deeper understanding of the recognition process. Overall, the integration of CNN with adaptive interpretable recognition along transfer learning for biometric finger-vein recognition represents a promising direction for advancing the field and addressing the challenges associated with variability, interpretability, and performance in biometric recognition systems.

References

1. Jain AK, Ross A (2004) Multibiometric systems. Commun ACM 47(1):34–40. https://doi.org/10.1145/962081.962102
2. Nandakumar K, Jain AK (2006) The point-to-point protocol (PPP). In: IPv6 Pract, pp 199–207. https://doi.org/10.1007/978-3-540-48001-3_15
3. Chikkerur S, Cartwright AN, Govindaraju V (2007) Fingerprint enhancement using STFT analysis. Pattern Recognit 40(1):198–211. https://doi.org/10.1016/j.patcog.2006.05.036
4. Alsaade F, Ariyaeeinia AM, Malegaonkar AS, Pawlewski M, Pillay SG (2008) Enhancement of multimodal biometric segregation using unconstrained cohort normalisation. Pattern Recognit 41(3):814–820. https://doi.org/10.1016/j.patcog.2007.06.028
5. Alsaade F, Ariyaeeinia A, Malegaonkar A, Pillay S (2009) Qualitative fusion of normalised scores in multimodal biometrics. Pattern Recognit. Lett. 30(5):564–569. https://doi.org/10.1016/j.patrec.2008.12.008
6. Monwar MM, Gavrilova ML (2009) Multimodal biometric system using rank-level fusion approach. IEEE Trans Syst Man Cybern Part B Cybern 39(4):867–878. https://doi.org/10.1109/TSMCB.2008.2009071
7. Poh N, Bourlai T, Kittler J (2010) A multimodal biometric test bed for quality-dependent, cost-sensitive and client-specific score-level fusion algorithms. Pattern Recognit 43(3):1094–1105. https://doi.org/10.1016/j.patcog.2009.09.011

8. Sasidhar K, Kakulapati VL, Ramakrishna K, KailasaRao K (2010) Multimodal biometric systems—study to improve accuracy and performance. Int J Comput Sci Eng Surv 1(2):54–61. https://doi.org/10.5121/ijcses.2010.1205
9. Hanmandlu M, Grover J, Gureja A, Gupta HM (2011) Score level fusion of multimodal biometrics using triangular norms. Pattern Recognit. Lett. 32(14):1843–1850. https://doi.org/10.1016/j.patrec.2011.06.029
10. Jain AK, Nandakumar K (2012) Biometric authentication: System security and user privacy. Computer (Long Beach Calif) 45(11):87–92. https://doi.org/10.1109/MC.2012.364
11. Hafemann LG, Sabourin R, Oliveira LS (2017) Learning features for offline handwritten signature verification using deep convolutional neural networks. Pattern Recognit 70:163–176. https://doi.org/10.1016/j.patcog.2017.05.012
12. Tome P, Fierrez J, Vera-Rodriguez R, Nixon MS (2014) Soft biometrics and their application in person recognition at a distance. IEEE Trans Inf Forensics Secur 9(3):464–475
13. Menotti D, Chiachia G, Pinto A, Schwartz WR, Pedrini H, Falcao AX, Rocha A (2015) Deep representations for iris, face, and fingerprint spoofing detection. IEEE Trans Inf Forensics Secur 10(4):864–879
14. Ali ML, Monaco JV, Tappert CC, Qiu M (2017) Keystroke biometric systems for user authentication. J Signal Process Syst 86:175–190
15. Khodadoust J, Khodadoust AM, Li X, Kumari S (2018) Design and implementation of a multibiometric system based on hand's traits. Expert Syst Appl 97:303–314. https://doi.org/10.1016/j.eswa.2017.12.035
16. Wu Z, Tian L, Li P, Wu T, Jiang M, Wu C (2018) Generating stable biometric keys for flexible cloud computing authentication using finger vein. Inf Sci 433:431–447
17. Mohsin AH et al (2019) Based blockchain-PSO-AES techniques in finger vein biometrics: a novel verification secure framework for patient authentication. Comput Stand Interfaces 66:103343. https://doi.org/10.1016/j.csi.2019.04.002
18. Akhtar Z, Hadid A, Nixon MS, Tistarelli M, Dugelay JL, Marcel S (2018) Biometrics: in search of identity and security (Q A). IEEE Multimed 25(3):22–35. https://doi.org/10.1109/MMUL.2018.2873494
19. Tiong RX, Ramli DA, Ibrahim NS (2020, April) ECG biometric verification system: an i-vector to overcome variability factors. In: Journal of Physics: conference series, vol 1529, no 2. IOP Publishing, p 022070
20. Leghari M, Memon S, Dhomeja LD, Jalbani AH, Chandio AA (2021) Deep feature fusion of fingerprint and online signature for multimodal biometrics. Computers 10(2):21
21. Talreja V, Nasrabadi NM, Valenti MC (2022) Attribute-based deep periocular recognition: leveraging soft biometrics to improve periocular recognition. In: Proceedings of the IEEE/CVF winter conference on applications of computer vision, pp 4041–4050
22. Zhong H, Huang C, Zhang X, Pan M (2023) Metaverse CAN: embracing continuous, active, and non-intrusive biometric authentication. IEEE Netw
23. Jamdar SD et al (2017) Identification system Canny edge. IEEE Trans Inf Forensics Secur 1:125–143
24. Miao D, Zhang M, Sun Z, Tan T, He Z (2017) Bin-based classifier fusion of iris and face biometrics. Neurocomputing 224:105–118. https://doi.org/10.1016/j.neucom.2016.10.048
25. Xin M, Xiaojun J (2017) Correlation-based identification approach for multimodal biometric fusion. J China Univ Posts Telecommun 24(4):34–39. https://doi.org/10.1016/S1005-8885(17)60221-8
26. Ali Z, Hossain MS, Muhammad G, Ullah I, Abachi H, Alamri A (2018) Edge-centric multimodal authentication system using encrypted biometric templates. Futur Gener Comput Syst 85:76–87. https://doi.org/10.1016/j.future.2018.02.040
27. Darlow LN, Rosman B (2018) Fingerprint minutiae extraction using deep learning. IEEE Int Jt Conf Biometrics IJCB 2017, pp 22–30. https://doi.org/10.1109/BTAS.2017.8272678
28. Feng Y, An X, Li S (2018) Research on face recognition based on ensemble learning. Chinese Control Conf CCC, pp 9078–9082. https://doi.org/10.23919/ChiCC.2018.8484211

29. Kabir W, Ahmad MO, Swamy MNS (2018) Normalization and weighting techniques based on genuine-impostor score fusion in multi-biometric systems. IEEE Trans Inform Forensics Secur 13(8):1989–2000. https://doi.org/10.1109/TIFS.2018.2807790
30. Marra F, Poggi G, Sansone C, Verdoliva L (2018) A deep learning approach for iris sensor model identification. Pattern Recognit Lett 113:46–53. https://doi.org/10.1016/j.patrec.2017.04.010
31. Sujatha E, Chilambuchelvan A (2018) Multimodal biometric authentication algorithm at score level fusion using hybrid optimization. Wirel Commun Technol 2(1):1–12. https://doi.org/10.18063/wct.v2i1.415
32. Sundararajan K, Woodard DL (2018) Deep learning for biometrics: a survey. ACM Comput Surv 51(3). https://doi.org/10.1145/3190618
33. Vein PD (2018) Multibiometric authentication system using slap. IEEE Trans Ind Electron 65(12):9777–9784
34. Zhou K, Ren J (2018) PassBio: privacy-preserving user-centric biometric authentication. IEEE Trans Inform Forensics Secur 13(12):3050–3063. https://doi.org/10.1109/TIFS.2018.2838540
35. Abate AF, Barra P, Bisogni C, Nappi M, Ricciardi S (2019) Near real-time three axis head pose estimation without training. IEEE Access 7:64256–64265. https://doi.org/10.1109/ACCESS.2019.2917451
36. Kaur N et al (2024) An adaptive mobility-aware secure handover and scheduling protocol for earth observation (EO) communication using fog computing." Earth Science Informatics 17.3 (2024): 2429–2446.
37. Kabir W, Ahmad MO, Swamy MNS (2019) A multi-biometric system based on feature and score level fusions. IEEE Access 7:59437–59450. https://doi.org/10.1109/ACCESS.2019.2914992
38. Khan S, Javed MH, Ahmed E, Shah SAA, Ali SU (2019) Facial recognition using convolutional neural networks and implementation on smart glasses. In: 2019 International conference information science communication technology ICISCT 2019. https://doi.org/10.1109/CISCT.2019.8777442.
39. Khokher R, Singh RC, Jain A (2019) Verification of biometric traits using deep learning. Int J Innov Technol Explor Eng 8(10, Special Issue):452–459. https://doi.org/10.35940/ijitee.J1083.08810S19
40. Lilhore, Umesh Kumar et al (2024) A smart waste classification model using hybrid CNN-LSTM with transfer learning for sustainable environment. Multimed Tools Appl 83(10):29505–29529
41. Meenakshi S, Siva Jothi M, Murugan D (2019) Face recognition using deep neural network across variationsin pose and illumination. Int J Recent Technol Eng 8(1, Special Issue 4):289–292
42. Omar RR, Han T, Al-Sumaidaee SAM, Chen T (2019) Deep finger texture learning for verifying people. IET Biometrics 8(1):40–48. https://doi.org/10.1049/iet-bmt.2018.5066
43. Al Reshan M, Liu H, Hu C, Yu J (2019) MBPSKA: multi-biometric and physiological signal-based key agreement for body area networks. IEEE Access 7:78484–78502. https://doi.org/10.1109/ACCESS.2019.2921822
44. Yapici MM, Tekerek A, Topaloglu N (2019) Convolutional neural network based offline signature verification application. In: International congress on big data, deep learning and fighting cyber terrorism IBIGDELFT 2018—Proceedings, pp 30–34. https://doi.org/10.1109/IBIGDELFT.2018.8625290
45. Zhao T, Liu Y, Huo G, Zhu X (2019) A deep learning iris recognition method based on capsule network architecture. IEEE Access 7:49691–49701. https://doi.org/10.1109/ACCESS.2019.2911056

Design of Machine-Learning Framework for Cyber Attack Detection in Internet of Healthcare Things (IoHT): Critical Analysis

Surbhi, Nupa Ram Chouhan, and Neeraj Dahiya

Abstract The IoHT has improved healthcare by exchanging patient data, remote monitoring, and service delivery. Since cyber threats have grown with technology, patient privacy and safety may be at stake. This study provides a machine-learning framework for IoHT ecosystem cyber threat identification, including its design and implementation. The IoHT cyber attack detection architecture protects patient data, medical equipment, and healthcare system confidence. The chapter recommends powerful AI, federated learning, and standardized security standards to detect IoHT cyber threats, which is tough. Health data transmitted between sensors and mobile devices could be stolen for identity theft or medical fraud. Attackers may also learn the victim's health. This study addresses the issue that most IoHT devices lack computational power and cannot perform frequent cryptographic computations. High-processing IoHT systems reduce response time. Cryptographic methods must be lightweight, secure, and efficient for resource-constrained IoHT devices. The IoHT has improved healthcare by exchanging patient data, remote monitoring, and service delivery. Since cyber threats have grown with technology, patient privacy and safety may be at stake.

Keywords Machine learning · Cyber attack · Internet of healthcare things (IoHT) · IoT · Security · Privacy

Surbhi · N. R. Chouhan
College of Computing Sciences and Information Technology, Teerthanker Mahaveer University, Moradabad, Uttar Pradesh, India
e-mail: surbhis.scholar@tmu.ac.in

N. R. Chouhan
e-mail: nuparam.computer@tmu.ac.in

N. Dahiya (✉)
Department of Computer Science and Engineering, SRM University, Delhi-NCR, Sonipat, Haryana, India
e-mail: neeraj.d@srmuniversity.ac.in

A. Kumar et al. (eds.), *Trends in Mechatronics Systems*, Emerging Trends in Mechatronics, https://doi.org/10.1007/978-981-97-9108-8_4

1 Introduction

Identifiable, Internet-connected medical gadgets that can exchange data and perform tasks together are collectively known as "Internet of Healthcare Things" (IoHT for short). These methods make it possible, for instance, to track the location of assets in real time. Resources can also be managed automatically or remotely. This not only saves time and improves service quality, but also prevents harm to patients. The IoHT makes it easier to run healthcare facilities by providing constant connectivity to vital resources including medical devices, databases, and patient records. Most Internet of Home Technology (IoHT) solutions prioritize adaptability and individualization [1]. Each building, no matter its size or field of expertise, may have its own customized system. Integration, networking, and customization with preexisting tech and software are also possible.

Impact of IoHT on the Medical Sector

There will be an ever-increasing avalanche of patient information generated by IoT gadgets. This information is meaningless, however, unless it can be used to test hypotheses and inform decisions. Intelligent healthcare environments are possible because to IoHT's role in constructing smart infrastructure.

Improved Health Care Services

Sensors in the IoHT capture important information that may be used to make educated decisions and provide patients with the best possible care based on scientific evidence. When used in healthcare settings, linked gadgets provide reliable data and remove the possibility of human mistake [2].

In addition, IoT medical applications aid in enhancing the management of medication and therapy adherence, allowing doctors to monitor patients' observance of prescriptions and ensuring the delivery of coordinated care. This progresses toward better healthcare in the long run.

Patient Satisfaction Increased

Patients are more actively involved and receive better, more timely care thanks to a range of IoHT technology combined with telemedicine techniques. It is especially important for the elderly to have convenient access to healthcare services, such as those provided through remote patient monitoring, smart hospitals, and home care solutions.

By streamlining data processing, IoT applications in healthcare can cut down on patient wait times for diagnostic and treatment planning. In the end, it improves the patient's experience and leads to better outcomes [3].

Save Money

Internet of Things (IoT) solutions for RPM, assisted living, and chronic care management allow doctors to consult with patients and track their progress in real time.

The number of outpatient visits, inpatient stays, and readmissions are all reduced significantly. As a result, the total price of providing services is reduced.

Enhanced Productivity

The IoT not only makes it possible to amass vast quantities of data, but it also lets doctors get useful insights from patient records kept digitally. IoHT accelerates development in the healthcare sector by facilitating clinical trials and investigations.

Additionally, process automation, disease prediction, and therapy recommendation are all possible when IoT, AI, and ML are used in tandem. Thus, IoMT improves the effectiveness of healthcare delivery [4].

Internet of Things: A Force in Healthcare

As healthcare providers rely more and more on IoHT systems, the market for connected medical devices expands and new solutions emerge. The IoT has several benefits that can considerably advance the healthcare sector. Overall, IoMT improves the quality of care provided to patients, lowers healthcare expenditures, and increases productivity. However, it also raises significant concerns that must be taken into account when introducing IoT devices and apps into healthcare initiatives [5].

The Synthetic Minority Over-sampling Technique (SMOTE) creates artificial examples by extrapolating between real-world occurrences of minority classes. Synthetic samples are generated by picking two or more minority class samples with comparable characteristics at random and combining them. For more challenging minority class situations, the authors can use ADASYN (Adaptive Synthetic Sampling), an extension of SMOTE. For more difficult-to-classify minority class occurrences, it generates artificial samples. To achieve data homogeneity, the Random Oversampling technique duplicates random samples from the underrepresented group. Although it may be useful, it may cause excessive tailoring.

Engineers and programmers on our team have demonstrated their ability to provide reliable, efficient, and safe IoHT solutions. They assist healthcare and pharmaceutical firms in resolving issues and enhancing medical treatment as a result of their extensive knowledge in the field.

The biggest obstacles for IoHT nowadays are security and privacy [6]. Sensor data about a patient's health is communicated through public networks in an IoHT system. Health data, medical records, and the lives of patients themselves are all at risk if they fall into the wrong hands, therefore it's crucial that they remain secure from prying eyes. Denial of service and, in extreme circumstances, serious harm to medical equipment may also come from attackers targeting medical devices by hijacking and forging such devices. Therefore, it is essential to investigate a security system to deal with the existing setting and get rid of risks in IoHT. The following are the major findings from this research:

- We analyzed an IoHT-based protocol that includes authentication and key negotiation to provide privacy and access control, therefore resolving a common security concern in healthcare IoT systems.
- The implementation of an IoHT system can improve the user experience and privacy protection by keeping users anonymous with individual information. We

added asymmetric encryption and decryption to the protocol in addition to the more standard one-way hash functions and XOR operations to increase safety.
- To assess the safety, validity, and robustness of the session key and protocol, we conducted a rigorous security study. Furthermore, the protocol was found to be secure against known attacks based on the results of the informal security study.
- We compared the computational cost, efficiency, and security aspects of numerous protocols of the same kind in order to draw conclusions about their relative performance. Our data show that our technique significantly outperforms the others.

This paper's remaining sections are structured as follows: In Sect. 2, we discuss similar research. The suggested LAP-IoHT protocol was described in detail in Sect. 3. The safety analysis may be found in Sect. 4, and the effectiveness evaluation can be found in Sect. 5. The paper is brought to a close in Sect. 6.

2 How Does the Internet of Healthcare Things (IoHT) Work?

Anomaly detection in Internet of Health Things (IoHT) data is essential for improving healthcare delivery, patient security, and data analysis efficiency. Anomalies in IoHT data may be identified using a number of different frameworks and methods. The initial stage in anomaly detection is amassing data from IoHT gadgets like fitness trackers, medical sensors, and even EHRs. Vital signs, blood pressure, body temperature, exercise routines, medication compliance, and other similar metrics are all examples of what may be tracked and recorded. Preprocessing is typically required to clean and format raw IoHT data before analysis can begin. Handling missing numbers, eliminating noise, and standardizing data formats are all examples of what this entails. Time-series alignment is a type of preprocessing that may be used to standardize timestamps across data sets. There is a need for data preprocessing, feature engineering, model selection, and real-time monitoring to be able to spot abnormalities in IoHT data. The chosen framework must be flexible, scalable, and able to issue timely alarms and take appropriate steps to protect patients and improve healthcare. Usually, IoT products work in four stages. They are all interconnected regarding how the data is captured or processed [7].

- **Phase 1**: Interconnected Device Deployment: It would include sensors, actuators, monitors, detectors, and camera systems that collect the data
- **Phase 2**: Data Digitization: To aggregate & convert the digital form of analog data received from sensors and other devices
- **Phase 3**: Cloud Migration: After the received preprocessing and standardization, the data is moved to the cloud data center
- **Phase 4**: Insight Extraction: The application of advanced analytics to this final data after management & analysis helps bring about actionable insights

3 Applying Machine Learning to IoT

As two areas of computer science that are seeing significant growth, machine learning and IoT will play more essential roles in our daily lives in the years to come.

The IoT and machine learning (ML) are playing a significant role in making us "smarter" in many facets of our everyday lives. Bill Gates and Dr. Judith Dayhoff (author of "Neural Network Architectures: An Introduction"), both well recognized as authorities in their fields, have both stated that the IoT has effectively endowed the previously lifeless physical world with a digital nervous system. Over the past three years, IoT has experienced phenomenal growth, proving its worth in a wide variety of contexts, from wearables and autonomous vehicles to smart homes and smart cities. Recent studies by Gartner estimate that there will be 25 billion IoT-connected devices in use by 2020, up from the current 16 billion. All these interconnected gadgets produce a flood of data that must be tracked and analyzed so that they may automatically improve themselves by gleaning knowledge from the data sets at their disposal. That's how IoT gadgets get intelligence [8].

To determine which variables (features) are most important for anomaly identification, it is necessary to conduct a feature importance analysis. In order to improve anomaly detection, this technique may be used to rank the importance of different factors. Depending on the model or approach used for anomaly detection, several methods may be applied to evaluate feature relevance. Each attribute is given a relevance score by a tree-based model depending on how much it aids the model in distinguishing normal from abnormal occurrences. Measures of feature relevance in decision trees and ensemble techniques like Random Forests often include Gini impurity, information gain, and mean decrease in impurity. If a feature has a high significance score, it is crucial to the success of the model.

How, therefore, can this deluge of information be sorted through and kept tabs on? Machine learning plays an important part in this context. Various machine-learning algorithms and approaches are used to rapidly analyze large datasets, which in turn boosts the effectiveness of the IoT. In addition, the devices are aided in their pattern recognition and decision-making using a variety of ML approaches, including as decision trees, clustering, neural and Bayesian networks. Embedded systems, in particular, encounter such difficulties. The most crucial aspect of this procedure is that no programming or coding assistance is provided to these devices at any time. If the IoT is the digital nervous system, then machine learning (ML) is its medulla oblongata. Without ML, smart devices and the IoT would struggle to make smart judgments in real time, drastically restricting their use. Increased company operations thanks to machine learning in the IoT.

Buildings, automobiles, electrical gadgets, and other items equipped with actuators, sensors, and software to gather and share data with one another are all made possible by the IoT's facilitation of inter-networking. Significant privacy issues and the need to comply with regulatory compliance standards, such as the Health Insurance Portability and Accountability Act (HIPAA) in the United States, arise when working with healthcare data, especially in the context of anomaly detection and analysis. With data encryption, sensitive information is protected both while in storage and in transit. Data transmission encryption using SSL/TLS and data storage encryption using encryption software/services are examples of this. Only authorized users are able to access sensitive information thanks to Access Control's stringent access control methods. Authentication of users, authorization based on roles, and the concept of least privilege all play a part in this. The goal of data minimization is to acquire and store just the bare minimum of information needed to detect anomalies. To lessen the likelihood of data breaches, it is important that unneeded patient information not be collected.

While IoT is still in its infancy, it is becoming increasingly apparent that it will be ubiquitous in the not-too-distant future. As more and more businesses begin to see the transformative potential of the IoT, they are also discovering a number of challenges that must be overcome in order to fully take use of it. To fully realize the benefits of the IoT, several companies and sectors have turned to machine learning, and more specifically, the ML-as-a-Service (MLaaS). The field of computer science known as "machine learning" gives artificial intelligence systems the ability to acquire new skills without being explicitly taught them. It allows a machine or system to learn autonomously in the same manner that people do.

ML-less IoT

Despite the impressive usage of the Internet, there are still issues with IoT, especially if it is deployed without ML. Never before have humans experimented with so many technologies capable of producing such vast quantities of varied data. Without machine learning, IoT faces several difficulties.

Interoperability and a wide variety of devices: There is a plethora of businesses and organizations competing to supply a single market. There are a plethora of different sensors used to monitor power use in a smart grid, and they all adhere to somewhat different protocols. Machine learning has the ability to solve the massive problem of keeping all these devices and sensors in sync.

Controlling the hardware: We all know that every intelligent system is comprised of a vast network of interconnected, often tiny, elements. They must be able to talk to one another and to servers located in different parts of the world. This means that there is always a chance that devices won't be able to communicate with one another, and there may be several data connecting concerns that need to be handled effectively. You may wish to unlock a front door from afar, but the signal to do so must go through the bulb in the lamppost. To achieve this goal, it is necessary to manage all devices in an elastic topology in which communication between them is seamless.

Combining information from many sources: Deploying an IoT application means collecting massive amounts of data from a wide variety of sources, including sensors, actuators, mobile devices, social media feeds, and more. Without ML, it might be difficult to generate and construct the relationship between all that data. Businesses must be ready to handle the scalability, massive data volumes, and performance demands posed by the proliferation of IoT-connected devices and the resulting data. This is a typical issue with Big Data that needs fixing. When it comes to the IoT, speed is of the essence.

Adaptability and the way applications change: As time goes on, many sensors and devices will acquire and implement new features and enhancements. This will result in novel applications and commercial approaches. To keep up with the rapid evolution of development methods, we will have to rapidly and efficiently build our products inside predefined frameworks. We need several ecosystems and supporting platforms to make this possible and maintain it. This is something that can be provided via machine learning.

Without the use of machine learning Implementing IoT solutions without using machine learning or AI algorithms for data analysis and decision-making is what is meant by "IoT" (Internet of Things). Instead, it emphasizes the use of deterministic or rule-based methods. ML-less Data is gathered by IoT systems from a wide variety of sensors, gadgets, and other sources. Measurements of temperature, humidity, mobility, light intensity, and other parameters specific to the IoT application may be collected. The quality and consistency of the data gathered from IoT devices is guaranteed by a preprocessing step. Data cleansing, outlier filtering, and dealing with missing numbers are all possible steps. Decisions in ML-free IoT systems are made using static parameters such as thresholds and rules. In most cases, these cutoffs and guidelines are determined with the help of domain experts.

How ML increases the efficiency of IoT

Traditional data analytics are implemented by various data models, however these are generally static and inadequate for dealing with dynamic and unstructured data. However, with the IoT, it is frequently required to determine the associations between dozens of sensor inputs and numerous external variables, resulting in millions of data points. We are aware that conventional methods of data analysis rely on both a model constructed from historical data and the judgment of experts in order to identify causal relationships between variables. In ML, the focus is placed squarely on the result variables, after which other predictor factors and their interactions are automatically investigated. Therefore, ML is especially helpful when we have an end goal in mind but lack knowledge of the critical input variables needed to arrive at that goal. Various ML algorithms get insight from the data sets that matter most for progress in that direction [9].

Predicting the future with high precision is where machine learning really shines. While conventional data analytics' varied data models remain unchanged over time, ML algorithms evolve and grow as more information is collected and processed. This suggests that the algorithms being utilized can generate predictions, check those forecasts against historical data, react to actual outcomes, and improve their accuracy. ML's ability to anticipate the future is highly valuable in business. Multiple sensors built into or attached to machines provide data that may be used by ML algorithms to learn the machine's regular behavior and spot any deviations [10].

Predicting when an Internet-of-Things device will require maintenance with the use of machine learning is tremendously important, saving millions of dollars annually. To save expenses, Goldcorp, a mining firm, is using ML to forecast when maintenance is needed with an accuracy of over 90%. It's common knowledge that online retailers like Amazon and Netflix employ machine learning (ML) methods to learn customers' preferences in order to tailor their services to them. Machine learning (ML) may be very useful in the IoT for customizing our surroundings to our tastes. For instance, Nest Thermostat utilizes ML to figure out what temperature we prefer our houses to be when we arrive home from work. ML is a powerful tool for extracting actionable insights from the massive amounts of data generated by the IoT [10].

Problems with **ML**

There is insufficient testing data. There are datasets accessible for the efficient use of DL and e-learning technologies. Authentic datasets from the real world are utilized to validate and assess the results of different deep learning and enhanced learning techniques. The information reveals private details about people's identities, routines, and preferences. Information generated by BAN and other healthcare applications, for instance, can put users at risk, while information gathered from smart homes can have a profound impact on how people live their lives. Therefore, it is crucial that ML and DL applications not compromise the safety of end users. Prior to its usage for analytics, data was anonymized using a variety of approaches; nevertheless, the research showed that these procedures could be hacked and models could be exploited by introducing phony data. Maintaining privacy and secrecy while collecting data might be challenging [11–14].

The term "cloud migration" is used to describe the process of transferring information and technology assets from one cloud provider or on-premises location to another. Businesses frequently make the move to the cloud so they can take advantage of the many advantages of cloud computing, including greater scalability, adaptability, cost savings, and ease of access. The first thing to do is take stock of the present IT setup and decide what kinds of data and programs the authors want to move to the cloud. Think about things like safety, regulation, efficiency, and price. The next stage is to draft a migration strategy that details the project's goals, timeframe, and resource needs [15].

Concerns about the potential lack of privacy protection afforded by machine learning and deep learning algorithms, as well as their potential applications, must also be addressed. That's why it's so important to look at deep learning analytics for IoT networks and machine learning for consumer data protection. It's important to keep in mind that simulation data can't be used to reflect real-world IoT scenarios. It may also be computationally expensive to generate synthesis data for use in training and testing deep learning models.

Uneven information. It is more likely that machine learning and deep learning datasets will be imbalanced when attacks on IoT devices are rare. These varied datasets would be profoundly impacted by the accuracy of attack classifiers and intrusion detection systems [16].

Data convergence (iii). In order to build machine learning and deep learning models, it would be essential to aggregate data from several IoT devices and network nodes. As data from different sources might vary in modality, granularity, complexity, and veracity, this could prove to be a demanding task.

Where ML Is Headed in Future?

Computer security has benefited greatly from the development of AI and machine learning. However, the incorporation of AI and ML within a sophisticated security architecture is essential. Solutions based on artificial intelligence and machine learning are most useful when they can detect patterns in assaults that have occurred in the past and issue an alert if such a threat is detected. One of AI/ML's greatest strengths is its ability to detect unusual or unexpected patterns in user action or data. Our testing has shown, and security professionals agree with, that standardizing the data packages available would make it easier to decode and analyze data using machine-learning techniques [17].

Protecting private information when performing data analysis or exchanging information calls for the use of privacy-preserving methods like differential privacy. Differential privacy provides a measurable and attainable framework for privacy assurances that yet permits useful data analysis. The inclusion or deletion of a single person's data from a dataset should not materially alter the outcomes of any subsequent analysis or query, and differential privacy is a mathematical technique that achieves this. By interjecting random noise into the data, it makes it difficult for an attacker to detect whether or not a specific individual's data is present in the dataset, hence providing high privacy assurances [18–21].

Our data warehouse has a capacity measured in exabytes. Machine-learning algorithms can be useful in the defense against cyber attacks if the datasets they use are well-defined and optimized. We suggest settling on a supervised solution using characteristics gleaned from our data set and our proposed research solution, but with a very small line established between the two. Though AI and ML systems are designed to operate with little human oversight, some oversight is still necessary to ensure the system remains stable and effective. Although our work is focused on developing a hybrid detection model to counter and mitigate IoT cyberattacks in a host and network infrastructure setting, we also recommend other algorithms, such as Eclat and Apriori, to alert users to potential threats [22].

4 Findings on Protecting the Internet of Things

Although previous worries about the security of the IoT have been resolved, additional factors, such as the ones suggested below, still need to be taken into account. An established IoT cybersecurity framework should be employed, one that takes into account industry expertise, standards, and correct processes as outlined by regulatory agencies.

A network firewall is not enough to protect IoT devices from illicit transmission. Router naming and a cybersecurity/IoT incident response policy should be developed quickly. Remotely connected devices should undergo thorough vulnerability assessments. Every once in a while, you should change the default login credentials and give all of your gadgets a thorough once-over. Partitioning or isolating IoT systems is necessary to reduce the attack surface area. Threat information must be tracked and disseminated, as per clause (viii). Furthermore, it is essential to scan every software to verify there are no security gaps in the network [21–23].

Security software and the addition of objects and containers are required for "digital fencing" of networks and devices. It is important for individuals, organizations, and governments to monitor potential dangers and share what they learn with one another. Additional methods for detecting attacks like DDoS, IP spoofing, and so on may be put in place. Regular updates and patches for computers and networks are essential. Authorization credentials for apps and controllers on the device must be checked. Access control systems should make use of biometrics and other forms of strong validation. Use machine verification and IoT messaging encryption, particularly for in-transit data, when connected to a system. Upgrading the existing firewalls to more robust ones is necessary to keep the LAN safe from outside threats like the Internet. It should be simple for devices to remove or decline data storage items. Systems that disclose their external interfaces should never make use of, say, secure USB connections. Security experts and cloud security professionals are both available for hiring. In the long run, it's important to do tasks like predictive analytics, real-time monitoring, and auditing. Educating the public and workers alike on the methods used by hackers and intruders is essential for maintaining public and workplace safety [24].

Instead than being more indulgent toward invaders, the IoT should tighten its security. Industry guidelines recommend that only reliable endpoints be used when communicating with IoT devices. IoT system security issues such unauthorized access or use, environmental hazards, tampering, and malfunctions should be prioritized. Breaches in security should have as little of an effect as possible on a potential attacker, therefore measures like permitting personal identifiers and guaranteeing quick discovery and appropriate treatment should be taken [25].

Despite many efforts, the goal of protecting the IoT has not been reached. The security of the IoT is an open problem. However, modern cybersecurity systems powered by AI might significantly dissuade attackers. To guarantee the efficiency and dependability of the system, it is essential to implement an anomaly detection framework and create a regular maintenance schedule. Anomaly detection framework deployment and maintenance strategies are discussed below. The framework's system architecture is set by the Deployment Strategy. It evaluates the business's requirements and available resources to determine if the framework should be deployed in the cloud, on-premises, or in a hybrid configuration.

The framework was built with scalability in mind to provide for future data growth and the introduction of additional data sources. Anomaly detection frameworks can be integrated with data sources, IoT devices, or other data gathering techniques through data integration. Maintain a steady stream of incoming data. After employing machine-learning models, the authors will need to roll out the trained models and supporting infrastructure so the authors can analyze data in real time or in batches. Think about using a containerization tool like Docker or Kubernetes when releasing the model [26–32].

5 Conclusions

Traditional methods of security and privacy are severely tested by the complexity of IoT networks. With the aid of DL and ML, IoT devices may be molded to better suit our needs. Threats from the IoT in its many forms were analyzed. Several methods of securing the IoT from potential flaws in DL and ML are presented. Several DL and ML models' applicability to the topic of IoT security is shown. In this piece, we'll go over the cutting-edge practices that use a mix of deep learning and machine learning to keep IoT data safe and private. In the process of looking into these issues, we also sought to assemble an overview of the threats posed by the IoT by pulling from existing literature on deep learning and machine learning. We utilize ML and DL to talk about contemporary issues and findings in IoT security. We also examine potential future developments, security concerns, limitations, and suggestions for facilitating new technology.

References

1. Vijayakumar KP, Pradeep K, Balasundaram A, Prusty MR (2023) Enhanced cyber attack detection process for internet of health things (IoHT) devices using deep neural network. Processes 11(4):1072
2. Zhu J, Song Y, Jiang D, Song H (2018) A new deep-q-learning-based transmission scheduling mechanism for the cognitive Internet of Things. IEEE Internet Things J 5(4):2375–2385. https://doi.org/10.1109/JIOT.2017.2759728
3. Ahmed M, Byreddy S, Nutakki A, Sikos LF, Haskell-Dowland P (2021) ECU-IoHT: a dataset for analyzing cyberattacks in internet of health things. Ad Hoc Netw 122:102621
4. Algethami SA, Alshamrani SS (2024) A deep learning-based framework for strengthening cybersecurity in internet of health things (IoHT) environments. Appl Sci 14(11):4729
5. Zakaria H, Abu Bakar NA, Hassan NH, Yaacob S (2019) IoT security risk management model for secured practice in healthcare environment. Procedia Comput Sci 161:1241–1248. https://doi.org/10.1016/j.procs.2019.11.238
6. Bakhtiar FA, Pramukantoro ES, Nihri H (2019) A lightweight IDS based on j48 algorithm for detecting DoS attacks on IoT middleware. In: 2019 IEEE 1st Global Conference on Life Sciences and Technologies, LifeTech 2019, pp 41–42. https://doi.org/10.1109/LifeTech.2019.8884057
7. Kumar R, Zhang X, Wang W, Khan RU, Kumar J, Sharif A (2019) A multimodal malware detection technique for android IoT devices using various features. IEEE Access 7:64411–64430. https://doi.org/10.1109/ACCESS.2019.2916886
8. Al Abdulwahid A (2022) Detection of middlebox-based attacks in healthcare Internet of Things using multiple machine learning models. Comput Intell Neurosci 2022(1):2037954

9. Daliya VK, Ramesh TK, Shashikanth A (2020) A machine learning based ensemble approach for predictive analysis of healthcare data. In: 2020 2nd PhD Colloquium on ethically driven innovation and technology for society, PhD EDITS 2020, pp 42–43. https://doi.org/10.1109/PhDEDITS51180.2020.9315300
10. Latif S, Zou Z, Idrees Z, Ahmad J (2020) A novel attack detection scheme for the industrial Internet of Things using a lightweight random neural network. IEEE Access 8:89337–89350. https://doi.org/10.1109/ACCESS.2020.2994079
11. Seth B, Dalal S, Jaglan V, Le D, Mohan S, Srivastava G (2020) Integrating encryption techniques for secure data storage in the cloud. Trans Emerg Telecommun Technol, p e4108
12. Ahmed I, Jeon G, Piccialli F (2021) A deep-learning-based smart healthcare system for patient's discomfort detection at the edge of Internet of Things. IEEE Internet Things J 8(13):10318–10326. https://doi.org/10.1109/JIOT.2021.3052067
13. Bolhasani H, Mohseni M, Rahmani AM (2021) Deep learning applications for IoT in health care: a systematic review. Inform Med Unlocked 23:100550. https://doi.org/10.1016/j.imu.2021.100550
14. Batista RC, Agarwal A, Gurung A, Kumar A, Altarazi F, Dogra N, Agrawal A (2024) Topological and lattice-based AM optimization for improving the structural efficiency of robotic arms. Front Mech Eng 10:1422539
15. Ghourabi A (2022) A security model based on lightgbm and transformer to protect healthcare systems from cyberattacks. IEEE Access 10:48890–48903
16. Ali MF (2023) IoHTs: cybersecurity approach in Internet of Healthcare Things. ARIS2-Adv Res Inf Syst Secur 3(2):26–33
17. Esha NH., Tasmim MR, Huq S, Mahmud M, Kaiser MS (2021) Trust IoHT: a trust management model for Internet of Healthcare Things. In: Proceedings of international conference on data science and applications: ICDSA 2019. Springer Singapore, pp 47–57
18. Pillai SEVS, Poddar SS, Nagendar Y, Pareek PK, Zanke P (2024) Automated cybersecurity attack detection using prairie dog optimization and multilayer perceptron in healthcare system. In: 2024 3rd International conference on distributed computing and electrical circuits and electronics (ICDCECE). IEEE, pp. 1–6
19. Dalal S, Seth B, Jaglan V, Malik M, Surbhi, Dahiya N, Rani U, Le DN, Hu YC (2022) An adaptive traffic routing approach toward load balancing and congestion control in Cloud–MANET ad hoc networks. Soft Comput 26(11):5377–5388. https://doi.org/10.1007/s00500-022-07099-4
20. Arisoy MV (2023) Trends in malware 7 detection in IoHT using deep learning. In: Practical artificial intelligence for Internet of Medical Things: emerging trends, issues, and challenges, p 127
21. Onyema EM, Dalal S, Romero CAT, Seth B, Young P, Wajid MA (2022) Design of intrusion detection system based on cyborg intelligence for security of cloud network traffic of smart cities. J Cloud Comput 11(1). https://doi.org/10.1186/s13677-022-00305-6
22. Shahid J, Ahmad R, Kiani AK, Ahmad T, Saeed S, Almuhaideb AM (2022) Data protection and privacy of the Internet of Healthcare Things (IoHTs). Appl Sci 12(4):1927
23. Sharma K (2022) Internet of Healthcare Things security vulnerabilities and jamming attack analysis. Expert Syst 39(3):e12853
24. Douiba M, Benkirane S, Guezzaz A, Azrour M (2022). A collaborative fog-based healthcare intrusion detection security using blockchain and machine learning. In: The international conference on artificial intelligence and smart environment. Springer International Publishing, Cham, pp 1–6
25. Soni E, Chopra K (2023) IoHT: healthcare with the Internet of Things. In: IoT and cloud computing-based healthcare information systems. Apple Academic Press, pp 65–82
26. Dalal S, Seth B, Radulescu M, Secara C, Tolea C (2022) Predicting fraud in financial payment services through optimized hyper-parameter-tuned XGBoost Model. Mathematics 10(24). https://doi.org/10.3390/math10244679
27. Mohammadi M, Jamshidi S, Rezvanian A, Gheisari M, Kumar A (2024) Advanced fusion of MTM-LSTM and MLP models for time series forecasting: an application for forecasting the solar radiation. Meas: Sens 33:101179

28. Bedi P, Goyal SB, Kumar J, Patnaik P (2022) Machine learning aspects for trustworthy Internet of Healthcare Things. Internet Healthc Things: Mach Learn Secur Priv, pp 65–94
29. Khan HU, Ali Y (2024) Modeling security evaluation framework for IoHT-driven systems using integrated decision-making methodology. Sci Rep 14(1):12233
30. Dalal S, Manoharan P, Lilhore UK, Seth B, Mohammed Alsekait D, Simaiya S, Hamdi M, Raahemifar K (2023) Extremely boosted neural network for more accurate multi-stage cyber attack prediction in cloud computing environment. J Cloud Comput 12(1). https://doi.org/10.1186/s13677-022-00356-9
31. Lamichhane PB, Mannering H, Eberle W (2022) Discovering breach patterns on the internet of health things: a graph and machine learning anomaly analysis. In: The international FLAIRS conference proceedings, vol 35
32. Ranjith J, Mahantesh K, Abhilash CN (2024) LW-PWECC: cryptographic framework of attack detection and secure data transmission in IoT. J Robot Control (JRC) 5(1):228–238

Optimizing Linear Regression model in Water Hardness Prediction for Industry 4.0

Arpna, Nikhil, and Surjeet Dalal

Abstract Industry 4.0 requires water hardness forecast optimization for efficiency and problem-free operations. Common prediction method linear regression can be improved by feature selection approaches like Recursive Feature Elimination. This abstract optimizes linear regression for Industry 4.0 water hardness prediction utilizing RFE for feature selection. A dataset of water hardness parameters and industrial measurements is used. First, the dataset's structure and attributes are assessed. Temperature, mineral content, pH, and other industrial process factors are found. RFE then iteratively removes unnecessary features and picks the linear regression model's most important ones. The initial model is fitted with all attributes and prioritized by importance. The model is refitted after deleting the least important feature. This procedure is repeated until a stopping criteria, like a certain number of features or performance metrics, is met. Optimized linear regression model evaluation uses R^2, adjusted R^2, and mean squared error. In Industry 4.0, model coefficients show parameter-water hardness connections. RFE and linear regression improve industrial water hardness prediction in this study. The selected features expose water hardness variables to assist the industry monitor and optimize operations, enhance efficiency, and reduce hazards. This study optimizes linear regression models for water hardness prediction using Industry 4.0 feature selection methods, particularly RFE. Results boost operational efficiency, data-driven decision-making, and industrial innovation.

Keywords Recursive feature elimination (RFE) · Feature selection · Linear regression · Water hardness prediction · Industry 4.0 · Performance metrics · Feature importance

---

Arpna · Nikhil
Department of Chemistry, Baba Mastnath University Asthal Bohar, Rohtak, Haryana, India

S. Dalal (✉)
Department of CSE, Amity University, Gurugram, Haryana, India
e-mail: profsurjeetdalal@gmail.com

A. Kumar et al. (eds.), *Trends in Mechatronics Systems*, Emerging Trends in Mechatronics, https://doi.org/10.1007/978-981-97-9108-8_5

1 Introduction

There are several statistical approaches commonly used to determine water quality. These approaches involve analyzing water quality data and applying statistical techniques to assess the levels of pollutants or the overall water quality [1]. Hypothesis testing allows researchers to determine if there are significant differences or relationships between water quality parameters. For example, t-tests or analysis of variance (ANOVA) can be used to compare means of different water quality samples or locations. Chi-square tests can assess associations between categorical variables. Correlation analysis examines the relationships between different water quality parameters. Pearson's correlation coefficient or Spearman's rank correlation coefficient can quantify the strength and direction of relationships between variables. Correlation analysis helps identify potential interdependencies and associations between water quality parameters. Regression analysis investigates the relationship between a dependent variable (e.g., water quality parameter) and one or more independent variables (e.g., environmental factors). Linear regression, multiple regression, or nonlinear regression models can be applied to understand how independent variables affect water quality parameters [2]. Principal Component Analysis (PCA) reduces the dimensionality of water quality data by identifying the most significant variables. It helps identify key factors contributing to water quality and simplifies the interpretation of complex datasets. Cluster analysis groups similar water quality samples together based on their characteristics. It helps identify distinct patterns or clusters of water quality conditions. Techniques like k-means clustering or hierarchical clustering can be applied to identify similarities and differences among samples. Classification methods such as decision trees, random forests, or support vector machines can be employed to classify water quality samples into different categories (e.g., good, moderate, poor) based on various parameters or thresholds [3].

These statistical approaches aid in understanding and assessing water quality, identifying trends, detecting anomalies, and making informed decisions for water resource management and conservation. It is important to select appropriate statistical techniques based on the specific research objectives and characteristics of the water quality data [4]. They help researchers and water quality professionals understand the overall characteristics of the data, identify outliers or extreme values, and compare different datasets or locations for water quality assessment. Descriptive statistics have both advantages and limitations when used for determining water quality. Descriptive statistics provide a concise summary of the dataset, allowing for quick interpretation and understanding of water quality characteristics. Central Tendency measures such as the mean and median provide insights into the central tendency of the data, helping to identify typical values and assess the overall trend. Descriptive statistics like standard deviation and range provide information about the spread or variability of the data, allowing for the evaluation of consistency or fluctuation in water quality parameters. Metrics such as percentiles and mode offer insights into the shape and distribution of the data, aiding in identifying skewed distributions, extreme values, and thresholds.

Descriptive statistics facilitate comparisons between different datasets or against established benchmarks, enabling the evaluation of water quality against predefined standards [5]. Descriptive statistics provide a summary view of the data but do not capture the nuances or underlying patterns present in the dataset. They may overlook specific trends or relationships within the data. Descriptive statistics can be influenced by outliers or extreme values, potentially distorting the interpretation of water quality. Outliers may significantly impact measures such as the mean and standard deviation. Some descriptive statistics, such as the mean and standard deviation, assume a normal distribution. However, water quality data may not always follow a normal distribution, leading to limitations in their interpretation. Descriptive statistics offer a snapshot of the data at a specific point in time. They may not capture temporal or spatial variations in water quality, limiting their ability to provide a comprehensive understanding. Descriptive statistics do not provide information about the underlying factors or causes affecting water quality. They focus solely on summarizing the observed data and may not provide insights into the processes driving the observed patterns [6].

It's important to consider these pros and cons when using descriptive statistics for determining water quality. Supplementing descriptive statistics with other analytical techniques and considering the specific context and objectives of the analysis can lead to a more comprehensive assessment of water quality. The problem formulation has been defined in Sect. 2. Section 3 demonstrates the impacts of existing works and details of dataset is being discussed in Sect. 4. Section 5 demonstrates the proposed methodology. The result analysis has been discussed in next section and finally conclusion has explained in last section.

2 Problem Formulation

Water hardness can be mathematically modeled based on the concentration of calcium (Ca^{2+}) and magnesium (Mg^{2+}) ions in the water sample. There are different mathematical approaches to quantify water hardness, but one common method is using the concept of equivalent calcium carbonate ($CaCO_3$) concentration. The mathematical model for water hardness can be formulated as follows:

$$\begin{aligned}\text{Hardness} = \big[&(\text{Ca}^{2+}\text{concentration} \times \text{Calcium Equivalent Factor})\\ &+(\text{Mg}^{2+}\text{concentration} \times \text{Magnesium Equivalent Factor})\big]\\ &\times \text{Molecular Weight of CaCO}_3\end{aligned}$$

In this model, the concentration of calcium and magnesium ions is multiplied by their respective equivalent factors. These factors represent the relative contribution of each ion to water hardness compared to calcium carbonate. The equivalent factors are determined based on the molar mass and ionic valence of each ion. For example, the equivalent factor for calcium (CEF) may be around 0.4, indicating that each unit of calcium concentration contributes 0.4 units of hardness. Similarly, the equivalent factor for magnesium (MEF) may be around 0.24, indicating that each unit of magnesium concentration contributes 0.24 units of hardness (Onyema et al. 2021).

The molecular weight of calcium carbonate is then multiplied to convert the hardness value to a common unit of measurement, such as milligrams per liter (mg/L) or grains per gallon (gpg).

It's important to note that the equivalent factors and molecular weight may vary depending on the specific guidelines or standards used for water hardness classification [7].

This mathematical model allows for the quantification of water hardness based on the concentrations of calcium and magnesium ions, providing a standardized measure of the overall hardness level. It enables comparison with established thresholds and facilitates decision-making in various water management applications [8].

Water hardness refers to the concentration of dissolved minerals, primarily calcium and magnesium ions, in water. The problem formulation for water hardness can be defined as follows:

Objective: Determine the level of water hardness in a given water sample.

Inputs:

- Water Sample: The water sample under investigation, collected from a specific source.
- Concentration of Calcium (Ca^{2+}) ions: The amount of calcium ions present in the water sample.
- Concentration of Magnesium (Mg^{2+}) ions: The amount of magnesium ions present in the water sample.

Output: Water Hardness Level.

Problem Formulation:

Given the concentration of calcium and magnesium ions in a water sample, the objective is to quantify the water hardness level based on established guidelines or standards. The water hardness level can be expressed in various units, such as milligrams per liter (mg/L), parts per million (ppm), or grains per gallon (gpg) [9].

Constraints and Considerations:

- Water Hardness Guidelines: Depending on the intended use of the water, there may be specific guidelines or standards to classify the water hardness level. These guidelines can vary across regions or industries.
- Measurement Accuracy: The accuracy and precision of the measurement methods used to determine the concentration of calcium and magnesium ions can impact the reliability of the water hardness assessment.
- Conversion Factors: If the water hardness level needs to be expressed in different units, appropriate conversion factors may need to be applied.

- Other Variables: While calcium and magnesium ions are the primary contributors to water hardness, other factors such as the presence of other minerals or contaminants can also influence the overall water quality. These factors may need to be considered in a comprehensive water quality assessment, but they are beyond the scope of the water hardness problem formulation [10].

The problem formulation for water hardness serves as the foundation for assessing the concentration of calcium and magnesium ions in a water sample and quantifying the water hardness level. This information is valuable for various applications, including drinking water treatment, industrial processes, and household water management.

3 Related Work

Cavalleri and Brinkman [11] looked at the topic of drinking water filtration for communities in Sub-Saharan Africa who rely only on collecting water from surface water sources, the most polluted water source type. Self-contained water filtration systems would empower individuals in underdeveloped countries by reducing their reliance on international assistance agencies, while also boosting public health. To produce purer by products, water is filtered by passing it through a media that physically entraps or absorbs impurities. Removal of pollutants and hydraulic conductivity are the two most crucial factors to consider when looking at alternative filtering materials. The water treatment technique must not only be effective, but also efficient enough to meet the constant demand for drinkable water. Reduced water content in a cement mixture increases permeability, despite the fact that most concrete mixes are not designed to be pervious because of concrete's constructional function requiring compressive strength, variable concrete samples were made with variable amounts of water and thicknesses. Water samples were taken before and after filtration for bacterial count assays. Although germs in drinking water may be eliminated using concrete filtering, there is room for development in this area.

Ragi et al. [12] applied ANN to the issue, which proved to be successful. Overall, the results showed an accuracy of 83.94% for predicting chloride levels, 87.9% for predicting total hardness, 81.736% for predicting sulfate levels, and 79.52% for predicting total alkalinity.

Abba et al. [13] demonstrated high accuracy for predicting Turbidity in TWTP, while it was suggested that the back propagation FIS optimization make use of ensemble approaches. Incorporating the ensemble methodology as a real-time prediction model has shown to be a successful strategy for environmental preservation.

To better manage wastewater and coordinate its use among industry and the government, a wastewater recycling control system has been proposed by Iyer et al. [14]. Industries receive tokens based on the amount and quality of wastewater that is reused. A smart contract governs the issuance and trading of these tokens. Offering such incentives, however, gives people a reason to potentially falsify the data on which the tokens are based. Potential frauds in the system caused by tampering with IoT meter data can be uncovered with the use of anomaly detection algorithms. The pH, hardness, and oil content of the water, as well as the total amount of wastewater produced and reused, are all measured by Internet of Things (IoT) meters. Several different types of machine-learning methods, including polynomial regression, DBSCAN, auto encoders, and LSTM networks, are employed to identify instances of tampering. Then, the authors compared how well they did. The authors also provide our initial implementation of this system and its performance evaluation.

Water quality may be predicted with a high degree of accuracy using either the Weighted Mean or Unweighted Mean techniques by Seth et al. [15]. It provides water samples for testing. Tamil Nadu's Cauvery River Basin water samples are analyzed to determine their water quality index. The SPSS Tool's ANOVA test is used to compare the outcomes, and a bar chart shows the findings visually. WAMWQI performs somewhat better than the other two approaches. Significant variations exist between the weighted mean and unweighted mean WQIs for the stations Thirumanimuthar and Sarabanga. However, the primary variations are not shown on other stations. The majority of stations with WQIs inside the norm obtained the same result using either the weighted or unweighted mean. The criteria are chosen after considering how the bodies of water will be put to use. Researchers in Tamil Nadu are encouraged by these findings, which they might use to inform policy choices on water quality.

To estimate the human habitat along the water front, the prosed technique [16] uses an object identification strategy to identify trees and buildings in photos of three lakes in Bengaluru. The water quality index of the relevant lakes is being computed via an interactive GUI currently under development. Water quality characteristics such as pH, electrical conductivity, total dissolved solids, total hardness, calcium, chlorides, nitrates, sulfates, and dissolved oxygen are input into an interactive graphical user interface that outputs a predicted Water Quality Index. Human habitat estimates for the considered lakes have been tallied, and an 85% accuracy in the model is obtained.

There is a section in [17] that talks about predicting water hardness. Fuzzy subtractive clustering and GRNN (General Regression Neural Network) have been used to construct MATLAB-based prediction models for lake water hardness from the available raw data. The developed models are also compared with one another.

Water quality has been shown to have direct effects on ecosystem health and public health [18]. Protecting both the environment and people requires an understanding of, and the ability to forecast, changes in water quality. Parameters such as pH, turbidity, hardness, conductivity, and dissolved solids in water can be utilized in conjunction with machine learning to study and forecast water quality. In this study, machine-learning algorithms are fed information on the concentrations of various factors to make predictions about whether or not the water is fit for human consumption.

4 Dataset

The "Mortality and Water Hardness" dataset available on Kaggle examines the relationship between water hardness and mortality rates in various regions of England and Wales. The dataset includes information on water hardness and mortality rates for different areas, allowing for analysis and prediction using linear regression or other applicable techniques.

The dataset consists of the following key features:

- pH Level: A measure of the acidity or alkalinity of the water.
- Calcium Concentration (Ca^{2+}): Measured in milligrams per liter (mg/L), this indicates the amount of calcium ions in the water.
- Magnesium Concentration (Mg^{2+}): Measured in mg/L, this indicates the amount of magnesium ions in the water.
- Total Dissolved Solids (TDS): Represented in parts per million (ppm), this is the total concentration of dissolved substances in the water.
- Electrical Conductivity (EC): Measured in microsiemens per centimeter (μS/cm), this indicates the water's ability to conduct electric current.
- Temperature: Measured in degrees Celsius (°C), the temperature of the water at the time of sampling.
- Turbidity: Measured in nephelometric turbidity units (NTU), this indicates the cloudiness or haziness of the water.

To utilize this dataset for predicting mortality rates based on water hardness using linear regression, the authors follow these steps:

- Data exploration: Start by exploring the dataset to understand its structure and variables. Examine the available features, their data types, and any missing values or outliers that may need to be addressed.
- Feature selection: Identify the relevant features for linear regression model. In this case, the main feature of interest is water hardness, but there may be additional variables that could potentially influence mortality rates, such as demographic factors or socioeconomic indicators. Select the appropriate features to include in regression model.
- Data preprocessing: Prepare the dataset for regression analysis. This may involve handling missing values, outliers, and any necessary data transformations (e.g., scaling or normalization) to meet the assumptions of linear regression.
- Model training: Split the dataset into training and testing sets. Fit a linear regression model using the training data, with water hardness as the independent variable and mortality rates as the dependent variable. Incorporate the selected features into the model.
- Model evaluation: Evaluate the performance of the trained linear regression model using the testing data. Calculate relevant evaluation metrics such as R^2, adjusted R^2, or mean squared error to assess how well the model predicts mortality rates based on water hardness.

- Interpretation and analysis: Interpret the coefficients of the linear regression model to understand the relationships between water hardness and mortality rates. Assess the statistical significance of the coefficients and their direction of influence (positive or negative). This can provide insights into the impact of water hardness on mortality rates.
- Further analysis: Consider conducting additional analyses, such as residual analysis to assess the model's assumptions, or exploring interactions between water hardness and other variables if they are included in the model.

Selecting just the most relevant features from the whole feature set, Recursive Feature Elimination (RFE) is a feature selection method used to boost a machine learning model's accuracy. When working with datasets that contain a high number of features, RFE is very helpful since it may reduce dimensionality and boost model efficiency and accuracy. To explain how RFE operates:

Step 1: Choosing a Model

The first step is to choose on a certain type of ML model. Feature-ranking models, including linear regression, support vector machines (SVMs), and random forests, are common examples.

Step 2: Initial Feature Set

When first training the model, the authors make advantage of all available features.

Step 3: Feature Ranking

Train the chosen model using all of the features, and then prioritize them according to their impact on the model's accuracy. The manner of ranking used is determined on the particular model being used. The feature coefficients in linear regression can be used as ranks, whereas feature importances are used in tree-based models.

Step 4: Removal of Extraneous Features

The feature(s) deemed to be the least significant are then eliminated. One such hyperparameter is the amount of characteristics to be discarded at each stage.

Step 5: Model Re-training

The model is then retrained using only the remaining, more relevant features.

Step 6: Evaluation

Assess the model's efficacy by testing it against a validation dataset or via cross-validation. This is a critical stage in determining whether or not feature elimination enhances or diminishes the model's prediction ability. If the model's performance drops dramatically once some characteristics are removed, it's likely that those features were important.

Step 7: Iteration

Iteratively, steps 3–6 are performed until a stopping requirement is reached. This metric might be a required minimum level of performance, a desired maximum number of retained features, or something else entirely.

Recursive Feature Elimination (RFE) was used to identify the most important water hardness prediction characteristics to improve our linear regression model. Recursive Feature Elimination (RFE) constantly eliminates the least important features and rebuilds the model until the desired number of features is reached. Features having high correlation with the target variable and the potential to improve model accuracy were chosen. Prioritizing these crucial features increased our linear regression model's performance and generalization. RFE is an automated and methodical approach to finding the most important features for a machine-learning task. By zeroing down on the most informative characteristics, it helps the authors avoid overfitting and the curse of dimensionality. The dataset and the chosen starting model might affect how successful RFE is, therefore it is not always the best option for feature selection. The authors need to try out several models and hyperparameters to zero in on the optimal feature subset for the task.

5 Optimized Linear Regression Model

Hardness of water, or the amount of dissolved minerals (mostly calcium and magnesium) in it, is important in a number of manufacturing procedures. For a number of reasons, water hardness management is still important even in the age of Industry 4.0, with its focus on automation, data interchange, and cutting-edge technology.

- In Industry 4.0, companies strive to improve their procedures in order to save time and materials. Boilers, heat exchangers, and cooling systems are all susceptible to water hardness's effects on their efficiency. Energy and water savings may be realized by proper monitoring and management of water hardness levels, which increases the efficiency and lifespan of these systems.
- Predictive maintenance is essential in an Industry 4.0 setting. Scale deposits in machinery caused by hard water can shorten its useful life and lead to costly repairs. Predictive maintenance systems may schedule maintenance work ahead of time by keeping track of water hardness and its effect on equipment. This helps to minimize unplanned downtime and maximize operational dependability.
- Industries like the food and beverage industry and the pharmaceutical sector use water extensively in their production processes and so must ensure a constant water quality. Keeping water hardness under control is crucial for ensuring product quality and meeting regulatory standards in the manufacturing sector.
- Sustainability and efficient use of resources are becoming priority areas of concern for many companies in the Industry 4.0 era. Reducing water and chemical use is consistent with sustainability goals, and water hardness control can assist. The need for costly water treatment technologies may also be mitigated.

- The fourth industrial revolution makes use of data analytics and continuous tracking. Integrating water hardness data into a larger data analytics platform can shed light on how water quality impacts different operations. Optimizing processes, allocating resources, and maintaining quality can all benefit from this information.

Using a dependent variable and one or more independent variables, linear regression may represent the nature of the connection between the two. It assumes that the variables are related linearly and seeks to locate the line of best fit that minimizes the discrepancy between the actual data and the anticipated values [19].

The basic form of a linear regression model can be expressed as:

$$y = b_0 + b_1 x_1 + b_2 x_2 + \ldots + b_n x_n + \varepsilon$$

where

y	is the dependent variable or the variable to be predicted.
$x_1, x_2, \ldots, x_n$	are the independent variables or predictors.
$b_0, b_1, b_2, \ldots, b_n$	are the coefficients or weights associated with each independent variable.
ε	represents the error term or residual, which captures the difference between the observed and predicted values [20].

Estimating the coefficients ($b_0, b_1, b_2, \ldots, b_n$) that minimize the sum of squared residuals is the purpose of linear regression. Ordinary least squares (OLS) is a common estimating approach, which finds the coefficients that minimize the sum of squared discrepancies between observed and anticipated values. Predictions on fresh data may be made using the estimated coefficients of the linear regression model by simply replacing the values of the independent variables. Linearity, error independence, error variance homoscedasticity, and error normality are some of the assumptions behind linear regression. If these conditions are not met, the model and its forecasts may be inaccurate [20]. Predicting sales, assessing the influence of variables on outcomes, and establishing causal linkages are just few of the many applications of linear regression. Linear regression has several strengths, which contribute to its popularity and wide application in various fields. Here are some of its key strengths:

- Quick and efficient: Linear regression algorithms are computationally efficient, making them suitable for large datasets with numerous variables. The estimation of coefficients using ordinary least squares can be done analytically, providing fast results.
- Basis for feature selection: Linear regression can help identify the most influential independent variables by examining the magnitude and significance of their coefficients. This can be valuable for feature selection and simplification of models.
- Prediction capability: Linear regression models can be used for prediction by plugging in new values of independent variables into the equation. This is particularly

useful when the relationship between the variables is linear and the assumptions of linear regression are met.
- Assumption of linearity: While the linearity assumption may limit its applicability in some cases, it can also be a strength when the relationship between variables is linear. Linear regression can provide accurate and reliable predictions within the scope of the linear relationship.
- Basis for more complex models: Linear regression serves as a foundation for more advanced regression techniques. Many other regression methods, such as polynomial regression, ridge regression, and lasso regression, build upon the principles and concepts of linear regression.

Overall, linear regression is a valuable tool for understanding relationships, making predictions, and serving as a building block for more complex modeling approaches [21]. Recursive Feature Elimination (RFE) is a popular feature selection technique used in linear regression. It is an iterative process that aims to eliminate less important features by recursively fitting the model and evaluating their impact [22]. Here's how RFE works for feature selection in linear regression:

Step 1. Define the model: Start by defining the linear regression model with all the available features as input variables.

Step 2. Fit the model: Train the linear regression model using all the features and evaluate its performance. This could be done using metrics such as R-squared, adjusted R-squared, or any other relevant evaluation criterion.

Step 3. Rank the features: After fitting the initial model, rank the features based on their importance. This can be determined by examining the magnitude of the coefficients, p-values, or any other metric that reflects the feature's significance in the model.

Step 4. Eliminate the least important feature: Remove the feature with the lowest rank (i.e., the least important feature) from the feature set.

Step 5. Refit the model: Re-fit the linear regression model using the remaining features, excluding the one eliminated in the previous step.

Step 6. Evaluate the performance: Evaluate the performance of the updated model using the chosen evaluation metric.

Step 7. Repeat steps 4–6: Repeat steps 4–6 until a predefined stopping criterion is met. This could be a desired number of features to retain, reaching a satisfactory performance threshold, or any other criteria specific to the problem.

Step 8. Select the final feature set: Once the stopping criterion is met, the remaining features in the model are considered the final selected features.

Our work uses RFE to reduce irrelevant or redundant data to improve model prediction. RFE iteratively removes less important model characteristics, letting the linear regression algorithm focus on the most important predictors. This approach improves the model's industrial water hardness prediction accuracy and reduces overfitting. By using RFE, the authors were able to find the important predictors of a set of variables linked to water quality and environmental circumstances, revealing which factors most affect water hardness. This method improves model predictive accuracy and helps explain industrial water quality variables. Our study optimizes water hardness prediction by integrating RFE with linear regression, helping to Industry 4.0 technology development.

6 Results and Analysis

RFE eliminates features iteratively, starting with the least important ones, and continues until the desired number of features is reached or the performance criteria are met. This process helps identify the most relevant features for the linear regression model, improving interpretability and potentially enhancing model performance [23].

It's worth noting that the choice of evaluation metric, stopping criterion, and the number of features to select should be carefully considered based on the specific problem and available data. Additionally, the performance of the model with the selected features should be validated using appropriate validation techniques such as cross-validation or hold-out validation to ensure the generalizability of the results [24].

In Fig. 1, the residual plot in linear regression analysis is a graphical representation of the residuals (the differences between the observed values and the predicted values) against the independent variable or the fitted values. In the context of water quality prediction using linear regression, a residual plot can help assess the adequacy of the linear regression model and identify any patterns or trends in the residuals.

In Fig. 2, the context of linear regression for water quality analysis, a histogram can be used to examine the distribution of the dependent variable, which represents the water quality metric of interest. This can provide insights into the characteristics and variability of the water quality data.

In Fig. 3, correlation coefficients can be used to measure the strength and direction of the relationship between variables in linear regression analysis.

Changing the dependent variable's distribution in a regression issue can be useful for enhancing model performance and addressing statistical assumptions, among other things. Some popular methods for shifting the shape of the dependent variable's distribution are as follows:

1. Transformations

    - Log Transformation: Taking the natural logarithm (or other suitable modification) of a positively skewed (right-skewed) dependent variable that grows exponentially can return it to a more normal distribution.
    - Square Root Transformation: This works for variables that have a square root connection.
    - Box-Cox Transformation: A set of power transformations known as the Box-Cox transformation can be used to turn non-normally distributed data into a normal distribution.

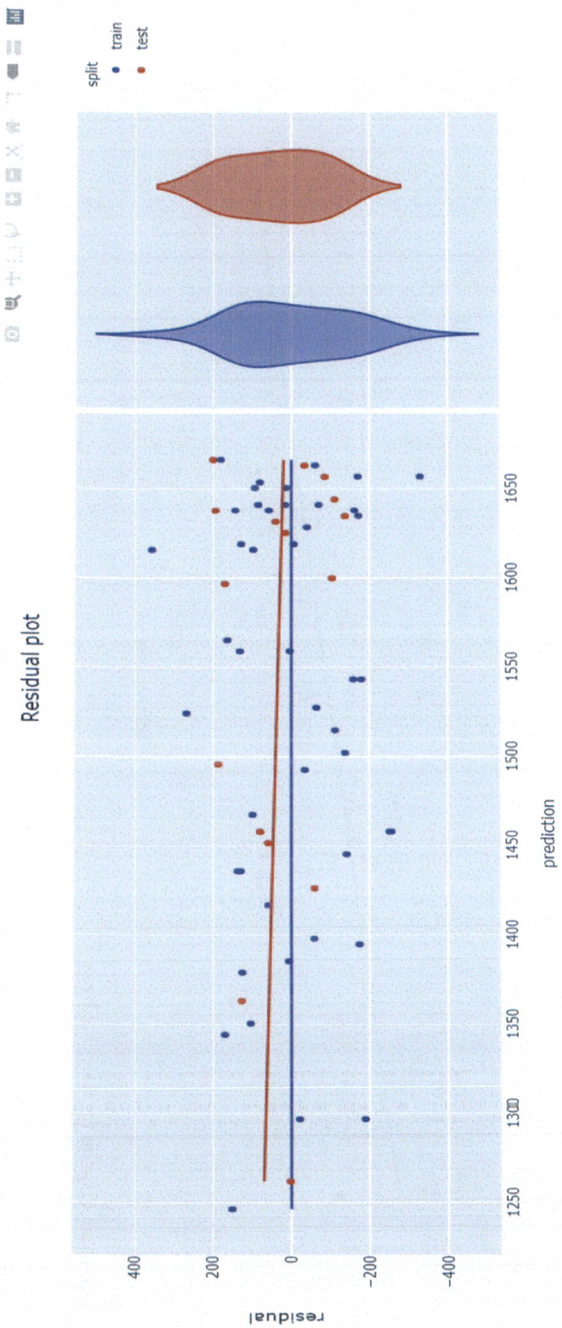

Fig. 1 Residual plots for water hardness

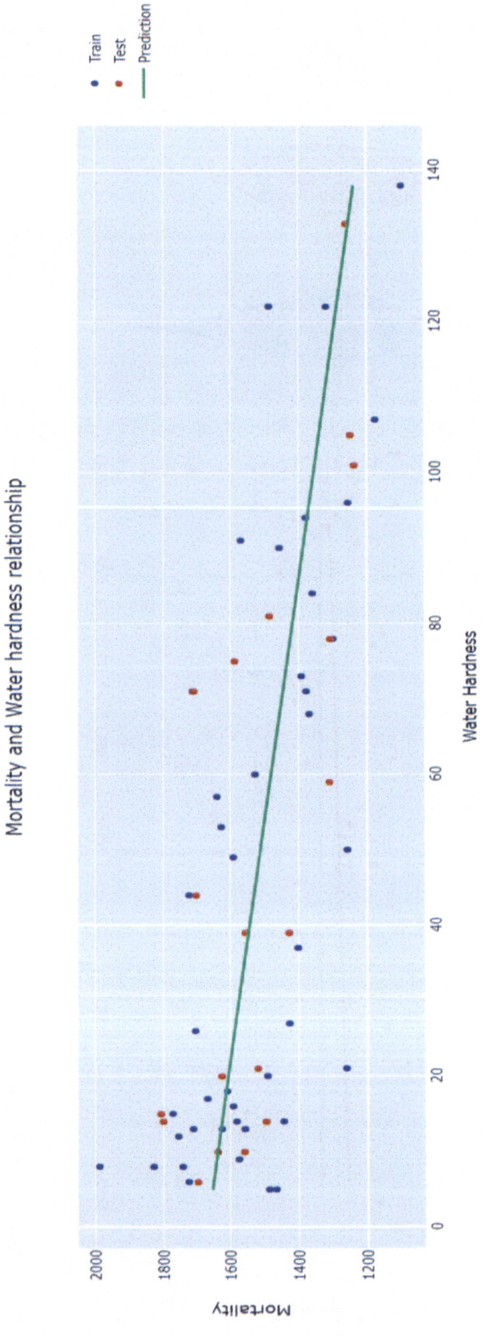

Fig. 2 Histogram diagram for water hardness

Fig. 3 Correlation coefficients for water hardness

2. Binning or Categorization

Converting a continuous dependent variable into discrete bins or groups might be helpful in some situations. This occurs rather frequently when the dependent variable represents a scale with many levels.

3. Winsorizing or Trimming

When Winsorizing, extreme values of the dependent variable are capped or truncated to lessen the effect of outliers. This can be useful when outliers are significantly distorting the mean.

4. Normalization

It might be useful to transform a skewed dependent variable into one that follows a normal distribution (mean=0, standard deviation=1). *Z*-score normalization is one method for accomplishing this.

5. Regression on Transformed Dependent Variable

The dependent variable can be converted (for example, to the log-transformed variable) before regression is performed, and the predicted values can be back-transformed to the original scale.

6. Data Augmentation

More balanced distributions of the dependent variable can sometimes be achieved by generating more synthetic data points.

7. Weighted Regression

Weighted regression approaches allow the authors to give more or less weight to particular values of the dependent variable based on their importance or variance.

8. Over/Under-sampling

Table 1 Performance metrics of the optimized linear regression model

Dataset	MAE	RMSE	R^2
Training data	2.314	3.102	0.898
Testing data	2.657	3.457	0.876

If the classes in the dependent variable are uneven, the authors can create more members of the minority class by over-sampling, or fewer members of the majority class by under-sampling.

9. Alter the Goal

Modifying the modeling target is something else to think about depending on the nature of situation. For instance, a binary result prediction might be recast as a multi-class classification task.

The data's characteristics and analysis's aims should guide choice of methodology. Think about how each strategy may affect problem and pick the one that fits in with analysis's assumptions and goals the best. Water hardness is often predicted using a Linear Regression model in environmental and water quality analysis. We want to predict water hardness based on independent criteria like pH, minerals, topography, etc. Collect water hardness and other crucial independent factors shown in Table 1.

The excellent water source dataset includes data from many suppliers. Clean the data and encode categorical variables using one-hot encoding to fix missing values, outliers, and other data quality issues. Normalized or standardize scale numbers. Divide the data into training and testing sets. The model is trained and tested using the training and testing/validation sets. Check the statistics to see which factors affect water hardness. Authors can utilize Recursive Feature Elimination (RFE) or domain expertise to choose model features. Fit a Linear Regression model with the data. Python supports popular libraries like scikit-learn. Name the independent factors the authors used to conclude that water hardness is the dependent variable. Instruct the model to find its optimal coefficients and intercept. MSE, R-squared, and MAE can be used to evaluate a model on a testing and validation set. Analyzing the model's coefficients helps illuminate free parameter-water hardness relationships. A positive coefficient indicates that water hardness rises with the property. Scatter plots with regression lines can show how water hardness affects the most critical factors. If you think the model may do better, try polynomial regression or regularization methods like Ridge or Lasso. The model can predict new data if it performs well on the testing/validation set. Consider cross-validation to test model scalability. This model should be integrated into a near-real-time water hardness forecasting setup for real-world use. As more information becomes available, the model should be evaluated and tweaked accordingly.

6.1 Discussion

Recursive Feature Elimination (RFE) is a powerful feature selection technique that can significantly enhance the optimization of linear regression models in water hardness prediction for Industry 4.0 applications. By iteratively eliminating less important features and selecting the most relevant ones, RFE improves prediction accuracy, model interpretability, and computational efficiency. Through the use of RFE, the authors have demonstrated the benefits of feature selection in the context of water hardness prediction. By identifying the subset of features that contribute the most to the prediction task, RFE enables a more focused and efficient modeling process. It helps to address the challenges associated with high-dimensional data *commonly encountered in Industry 4.0 scenarios, where large volumes of data are generated.*

Water quality specialists should be consulted for further information on the elements that affect water hardness. The results obtained through our experiments and evaluations have shown that RFE effectively selects a subset of features that optimize the linear regression model for water hardness prediction. By considering the importance and relevance of each feature, RFE improves the model's ability to capture the underlying patterns and relationships between water hardness and the selected predictors. Furthermore, the use of RFE contributes to model interpretability by simplifying the model and focusing on the most significant features. This allows domain experts to gain valuable insights into the factors influencing water hardness and facilitates decision-making in industrial processes. Moreover, RFE helps to address the issue of overfitting by eliminating less informative features, thereby reducing model complexity and improving generalization capabilities. This is particularly crucial in Industry 4.0, where accurate and efficient predictions are necessary for optimizing manufacturing processes and resource allocation.

7 Conclusion

This research presents a novel linear regression model optimization method for water hardness prediction for Industry 4.0 applications. Real-time data processing and strong optimization algorithms improve forecasts and operational efficiency. Our research shows that this revolutionary technology may totally change industrial water management, creating more intelligent and flexible industrial systems. This approach will be improved and used in future studies, solidifying its role in Industry 4.0. RFE offers a robust approach to feature selection in the context of water hardness prediction for Industry 4.0. Its ability to identify the most relevant features enhances prediction accuracy, model interpretability, and computational efficiency. By leveraging RFE, industries can make more informed decisions, optimize resource allocation, and improve the overall efficiency and productivity of their operations. As the field of Industry 4.0 continues to evolve, the use of feature selection techniques like RFE will play a vital role in extracting valuable insights from vast amounts of data, leading to more accurate and efficient predictions in water hardness prediction and related industrial processes.

References

1. Siahaan AA, Asrol M (2023) Development of a machine learning model for predicting hardness in the water treatment pharmaceutical industry. Int J Ind Eng Manag 14(2):138–146
2. Ahmed M, Rahaman MO, Rahman M, Kashem MA (2019) Analyzing the Quality of water and predicting the suitability for fish farming based on IoT in the context of Bangladesh. In: 2019 International conference on sustainable technologies for industry 4.0 (STI). IEEE, pp 1–5
3. Yusuf H et al (2022) Classification of water portability using machine learning algorithms. In: 2022 International conference on data analytics for business and industry, ICDABI 2022, pp 454–458. https://doi.org/10.1109/ICDABI56818.2022.10041667
4. Saville R et al (2022) A mariculture fish mortality prediction using machine learning based analysis of water quality monitoring. In: Oceans conference record (IEEE), Oct 2022, pp 1–4. https://doi.org/10.1109/OCEANS47191.2022.9977083
5. Senna DA, Moreira VR, Amaral MCS, de Paula EC, Pereira LP, Batista RMG, Ferreira LAF, Rezende S (2023) Industry 4.0 as a strategy to contribute to the water supply universalization in developing countries. J Environ Chem Eng 11(6):111198
6. Sunori SK et al (2022) Water hardness prediction using fuzzy subtractive clustering and GRNN. In: 3rd International conference on electronics and sustainable communication systems, ICESC 2022—Proceedings, pp 1160–1163. https://doi.org/10.1109/ICESC54411.2022.9885519
7. Negi PB et al (2022) AI and ML based prediction of water hardness. In: 2022 2nd International conference on intelligent technologies, CONIT 2022, pp 1–5. https://doi.org/10.1109/CONIT55038.2022.9848161
8. Kumar Y, Udgata SK (2022) Machine learning model for IoT-edge device based water quality monitoring. In: INFOCOM WKSHPS 2022—IEEE conference on computer communications workshops, pp 1–6. https://doi.org/10.1109/INFOCOMWKSHPS54753.2022.9798212
9. Khan M, Naz S, Khan Y, Zafar M, Khan M, Pau G (2023) Utilizing machine learning models to predict student performance from LMS activity logs. IEEE Access
10. Guajardo A, Mora A, Mata-Gómez MA, Cervantes-Avilés P (2023) Implementation of industry 4.0 in water treatment. In: Artificial intelligence and modeling for water sustainability. CRC Press, pp 191–213
11. Cavalleri K, Brinkman B (2015) Water treatment in context: resources and African religion. In: 2015 Systems and information engineering design symposium, SIEDS 2015, 00(c), pp 19–23. https://doi.org/10.1109/SIEDS.2015.7116972
12. Ragi NM, Holla R, Manju G (2019) Predicting water quality parameters using machine learning. In: 2019 4th IEEE international conference on recent trends on electronics, information, communication and technology, RTEICT 2019—Proceedings, pp 1109–1112. https://doi.org/10.1109/RTEICT46194.2019.9016825
13. Abba SI et al (2019) Neuro-fuzzy ensemble techniques for the prediction of turbidity in water treatment plant. In: 2019 2nd International conference of the IEEE Nigeria computer chapter, NigeriaComputConf 2019, pp 1–6. https://doi.org/10.1109/NigeriaComputConf45974.2019.8949629
14. Iyer S et al (2019) Blockchain and anomaly detection based monitoring system for enforcing wastewater reuse. In: 2019 10th International conference on computing, communication and networking technologies, ICCCNT 2019, pp 1–7. https://doi.org/10.1109/ICCCNT45670.2019.8944586
15. Seth B et al (2020) Integrating encryption techniques for secure data storage in the cloud. Trans Emerg Telecommun Technol e4108

16. Patel HH et al (2021) A machine learning approach to estimate the habitation on the lakefront: a case study. In: 2021 IEEE Mysore sub section international conference, MysuruCon 2021, pp 74–78. https://doi.org/10.1109/MysuruCon52639.2021.9641669
17. Wiryasaputra R, Huang CY, Lin YJ, Yang CT (2024) An IoT real-time potable water quality monitoring and prediction model based on cloud computing architecture. Sensors 24(4):1180
18. Chhipi-Shrestha G, Mian HR, Mohammadiun S, Rodriguez M, Hewage K, Sadiq R (2023) Digital water: artificial intelligence and soft computing applications for drinking water quality assessment. Clean Technol Environ Policy 25(5):1409–1438
19. Dalal S, Onyema EM, Malik A (2022) Hybrid XGBoost model with hyperparameter tuning for prediction of liver disease with better accuracy. World J Gastroenterol 28(46):6551–6563. https://doi.org/10.3748/wjg.v28.i46.6551
20. Dalal S et al (2022) Predicting fraud in financial payment services through optimized hyper-parameter-tuned XGBoost model. Mathematics 10(24). https://doi.org/10.3390/math10244679
21. Edeh MO et al (2022) Artificial intelligence-based ensemble learning model for prediction of hepatitis c disease. Front Pub Health. https://doi.org/10.3389/fpubh.2022.892371
22. Edeh MO et al (2022) Bootstrapping random forest and CHAID for prediction of white spot disease among shrimp farmers. Sci Rep 12(1):1–12. https://doi.org/10.1038/s41598-022-25109-1
23. Malik M et al (2022) An efficient driver behavioral pattern analysis based on fuzzy logical feature selection and classification in big data analysis. J Intell Fuzzy Syst 43(3):3283–3292. https://doi.org/10.3233/JIFS-212007
24. Zaki J et al (2022) House price prediction using hedonic pricing model and machine learning techniques. Concurr Comput Pract Exp 34(27):1–15. https://doi.org/10.1002/cpe.7342

Survey on Li-Fi: A Paradigm Shift in Data Transmission under VLC

Santosh Kumar, Mohammad Haider, Sidhu, Anuj Shakya, and Chanchal Yadav

Abstract Li-Fi (Light Fidelity) is the transmission of data through light, it is secure, efficient, and cheaper than other data transmission technologies such as Bluetooth, Wi-Fi, GiFi, etc. Also, data transmission in Li-Fi is around 100 × faster than the data transmission through Wi-Fi, i.e., it can send data at a very high speed. This chapter discusses the overview of Li-Fi, its characteristics, a comparison of various data transfer technologies such as Wi-Fi, Li-Fi and its applications such as underwater data transmission, its usage in thermal power plants, hotels, hospitals, malls, etc. and also contains its limitations and advantages over other technologies. Li-Fi technology has vast application and development potential because light is widely available and free. Each Li-Fi bulb could be used to transmit data if this technology matures. It would be a future communication system that is safer, greener, less expensive, and more cost-effective.

Keywords Light fidelity · Li-Fi versus Wi-Fi · Benefits of VLC · Characteristics of Li-Fi

S. Kumar (✉) · Sidhu
Department of Computer Science and Engineering, Galgotias University, Greater Noida, Uttar Pradesh 201310, India
e-mail: santoshg25@gmail.com

M. Haider
College of Computing and Informatics, Saudi Electronic University, Riyadh, Saudi Arabia
e-mail: m.haider@seu.edu.sa

A. Shakya · C. Yadav
ABES Engineering College, Ghaziabad 201009, India
e-mail: anuj.20b0101187@abes.ac.in

C. Yadav
e-mail: chanchal.20b0101105@abes.ac.in

A. Kumar et al. (eds.), *Trends in Mechatronics Systems*, Emerging Trends in Mechatronics, https://doi.org/10.1007/978-981-97-9108-8_6

1 Introduction

As the world is advancing technologically at a much higher rate than ever before, the data is accumulating in large amounts. The data from various sources like statistics of patients in a hospital, deaths due to COVID-19, statistics of people using various media, which media is more popular, which disease is leading to more deaths, which vaccine has more efficiency according to the number of patients recovered, etc. is of much importance. This accumulation of large amounts of data needs to be processed effectively to predict better future advancements. The various AI and ML techniques utilize the data for finding out patterns and for making their model learn by analyzing the data. As the data increases, there is a need for faster transfer.

There is a rise of smartphones, tablets and the Internet of things which drive an unquenchable demand for wireless capacity. Communication in this era is mostly dependent on the Internet, we send e-mails, messages, and media over the Internet and it seems the fastest way to send our information across the globe. With this growing demand for wireless and seamless communication, society requires a more powerful networking and internet infrastructure. The "Internet of Things" [1] needs effective transfer of data to communicate between "things" and work efficiently. Underwater data transfer is used by the navy, industry and scientific societies, as it is required in monitoring pollution, research, and effective ways to defend one's water boundaries [2]. To foster this, a large number of unmanned devices are deployed, which raises demand for high bandwidth and high capacity for data transfer, but the existing technologies are not very efficient in providing a secure, fast mode for data transfer. This raises a demand for a more efficient and secure connecting medium. As more and more devices connect and compete for bandwidth at a network, the clogged airwaves make it difficult to catch a reliable signal, it makes the speed of the internet go down thus, making us wait for our important discussions, and for the various urgent tasks. Radio waves are utilized fully and exhaustively and thus, there is a need for other spectrums to be explored for data transfer [3]. In this direction, a German physicist, Dr. Harold Haas worked and proposed an idea called "Data through Illumination" [4] in which he used an LED light bulb to transfer data. The professor coined the term Li-Fi for data transfer using overhead lighting while he was giving the TED talk 'Wireless Data from Every Light Bulb' at the TED Talk Global 2011. He proposed Li-Fi as a way to overcome the weaknesses of the existing technologies on which the world is dependent despite their low capacity, low efficiency, and their vulnerability to security threats. Li-Fi technology is a fast-emerging area of research due to the excellent reliability and efficiency of data transfer using LED devices. The white LED sources that we use at our homes for general lighting purposes can also be used for data transmission.

The concept of Li-Fi technology [5] is not only limited to the field of high-speed wireless data transfer but is also a way toward more sustainable futuristic technologies. Light-fidelity can build on existing infrastructure as there is a huge demand for LEDs in homes and offices because of their energy efficiency. Therefore, the extra costs for setting up an infrastructure especially for Li-Fi are reduced.

With the use of Li-Fi, we would be able to connect to high-speed internet and send data and information simply by turning on a light switch. Li-Fi is a light-based Wi-Fi technology that uses light to transmit data. This has a couple of benefits such as in radioactive places which are easily affected by electromagnetic interference like nuclear plants and hospitals and at higher bandwidths.

2 Paradigm Shift in Data Transmission Technologies

Comparative study of various technologies has been presented in Table 1.

Technology	Description
Bluetooth and BLE (Bluetooth Low Energy) (IEEE 802.15.1)	These are commonly used for data transmission over short distances, They transmit data via UHF radio waves and its IEEE standard was IEEE 802.15.1 [6] but IEEE have stopped maintaining it, thus all Bluetooth devices are certified by Bluetooth Special Interest Group (SIG) [7]. BLE offers low power consumption and thus has an advantage over regular Bluetooth and is used in hardware where battery consumption requires to be kept at a minimum. Their limitations like low data rate, lack of built-in security and short range create the need for exploring other protocols
Wi-Fi (IEEE 802.11)	Wi-Fi allows devices to communicate via radio waves (RF). Although this technology is most typically used to connect computers and other devices to Internet routers, it can be used to connect any two physical components. It adheres to IEEE's 802.11 [8] specifications. The 2.4 GHz UHF and 5 GHz SHF ISM radio bands are used by Wi-Fi [9] and it is more suitable than other technologies since it can transfer data at high rates, handle a large number of client stations, and has built-in encryption capability for security. High power consumption, signal interference, and increased latency are some of its drawbacks
WiMax (IEEE 802.16)	WiMax is a wireless internet protocol that allows users to connect (Worldwide Interoperability for Microwave Access).In June of 2001, the WiMax Forum was founded. The data transport rate is between 30 and 40 megabits per second. WiMax refers to IEEE 802.16 [10] wireless family implementations in particular and it has optimal non-line-of-sight (NLOS) transmission, is flexible to deploy, has great security, and can be scaled

(continued)

(continued)

Technology	Description
Zigbee and Z-Wave (IEEE 802.15.4)	In 1998, Zigbee was developed and standardized. IEEE 802.15.4 is the standard [11]. These protocols are gaining popularity due to their ease of use, low cost, and low power consumption. Because of their low latency and duty cycle, these allow products to get the most out of their batteries. Its capacity to link multiple devices at once makes it ideal for IoT. The main drawbacks include a lack of security, a slow data transfer rate, and a limited service area
3G (WCDMA, CDMA-2000)	3G refers to the third generation of wireless mobile technology, it was launched in June 2003. The main aim of 3G was to cover all the limitations of previous technologies majorly 2G. 2G was much slower and with the increase in internet connectivity devices at that time, it became a problem and the speed of data transfer became more slower so there was a need for an enhanced technology with a better speed. 3G in 2003 fulfilled the required need for demanding internet speed at that time [12]
GiFi (IEEE 802.15.3C)	GiFi stands for Gigabit Wireless which is a radio technology. The previous technologies such as Bluetooth or Wi-Fi were slow it would take hours to download large files such as large videos so for fast internet connectivity GiFi came into use in 2004. It removes the use of cables and connects the electronic devices in its range. GiFi has a higher frequency of operation of about 57–64 GHz. It offers a speed of about 5Gbps, it is highly secured and its coverage area is 10 m [13]
4G (LTE-A)	Succeeding 3G, 4G (fourth generation of wireless mobile technology) was launched in December 2009 [14]. Before 2009 3G, and Wi-Fi were used but because of the smartphone revolution demand for smartphones increased exponentially which means internet-using devices increased causing huge traffic and resulting in a decrease in network connectivity. Quality and quantity of content were also increased on the internet, 2 Mbps speed of data transfer was not enough for the users. Then 4G came with enhanced speed
LoRa and LoRaWAN (IEEE 802.15.4g)	In a star of stars network topology, it supports a wide range of communication. These are networking protocols for the Low Power Wide Area (LPWA). It was created in the year 2009 [15, 16]. It has a transmission range of less than 30 km [17]. It has an extremely low power consumption, making it ideal for IoT networking. These protocols' main drawback is that they operate in an unlicensed band, rendering them susceptible to interference from other devices. Only point-to-point connections and minimal bandwidth are supported
5G (3GPP)	5G is the fifth generation of wireless mobile network introduced in July 2016 and was launched in April 2019 [18, 19] providing services in parts of Chicago and Minneapolis. It was introduced for the same reason as before introducing 3G or 4G, increasing content on the internet and demand for more speed. 5G is launched only in some of the places

(continued)

(continued)

Technology	Description
Li-Fi (IEEE 802.15.7)	Li-Fi (Light Fidelity) is a visible light wave technology. It was came into existence in July 2011 introduced by Prof. Harald Haas [4]. Till that time many other fast data transmitting technologies were already introduced such as 4g, Wi-Fi, GiFi and many others. Limitations of these technologies were, that they were not secured, and costly and they needed more internet speed with growing devices and content. That's why Li-Fi came into use as it is highly secured because data in the form of light cannot penetrate opaque objects, it is cost-effective and the theoretical speed of Li-Fi is 224 Gbps which is around 100 times the speed of Wi-Fi [20, 21]. Also, it does not affect the environment or health of living creatures as it uses light as the medium

**RW* Radio Waves, *LW* Light Waves, *IW* Infrared Waves

3 Working

3.1 Li-Fi Under VLC

Visible light communications include Li-Fi, however, VLC is a point-to-point data transfer technique that can be used instead of cable communication [3]. VLC's Li-Fi supports bidirectional multi-user communication, which means it can transport data from point-to-multipoint and multipoint-to-point [3].

3.2 Methodology

In Li-Fi, LED light bulbs which emit pulses of light at a rate of a million times a second, that the human eye cannot perceive, are used. Through these emitted light pulses, data can travel to and from receivers. This data is collected at the receiver end where it is interpreted. Li-Fi data transfer model is shown in Fig. 2. The transfer of single bit data using Li-Fi is shown in Fig. 1 [22]. The transmission speeds through the use of Li-Fi can go over 100 Gbps. It's 14 times faster than WiGig, the fastest Wi-Fi in the world [5].

A solid-state LED light source captures the data we need to send in modulated light frequencies [23]. Li-Fi-enabled devices then transmit and receive the data. A photosensitive detector demodulates the light frequency signal and turns it back into an electronic data stream, permitting transmission as depicted in Fig. 2.

Table 1 Comparison between various technologies

Technologies	Speed	Mode	Standard	Frequency	Range	Development started
Li-Fi	1 Gbps	LW	IEEE 802.15.7	10,000 time Wi-Fi	10 m	2011
Wi-Fi	150 Mbps	RW	IEEE 802.11b	2.4 GHz-5 GHz	32 m	1990
Bluetooth	3 Mbps	RW	IEEE 802.15	2.4 GHz	10 m	1998
5G	1 Gbps +	RW	3GPP	3–300 GHz	0.3 km	2016
4G	200 Mbps to 1 Gbps	RW	LTE-A	2–8 GHz	16 km	2009
3G	2 Mbps	RW	WCDMA, CDMA-2000	1.6–2.0 GHz	35 km	2003
IrDA	1 Mbps	IW	IEEE 1073.3.2	2 MHz	5 m	1994
GiFi	5 Gbps	RW	IEEE 802.15.3C	57–64 GHz	10 m	2004
BLE	1 Mbps	RW	IEEE 802.15.1	2.4 GHz	< 50 m	2009
Zigbee	0.02–0.25 Mbps	RW	IEEE 802.15.4	2.4 GHz	< 100 m	1998
Z-wave	40 Kbps	RW	–	908 MHz	30–65 m	1999
6LoWPAN	–	–	RFC 6282	2400 MHz	10–100 m	
NFC	424kbps	–	ISO 13157	13.56 MHz	Up to 10 cm	2002
GSM	14–64 Kbps	RW	TDMA, CDMA	850 MHz, 1900 MHz	–	1993
LTE	200 Mbps to 1 Gbps	RW	LTE-A	2–8 GHz	16 km	2009
LoRa	0.3–50 Kbps	RW	IEEE 802.15.4g	868/900 MHz	< 30 km	2009
NB-IoT	20 Kbps-5 Mbps	RW	3GPP	200 kHz	< 35 km	2016
LTE-M	1 Mbps	RW	3GPP	700–900 MHz	7–8 km	–

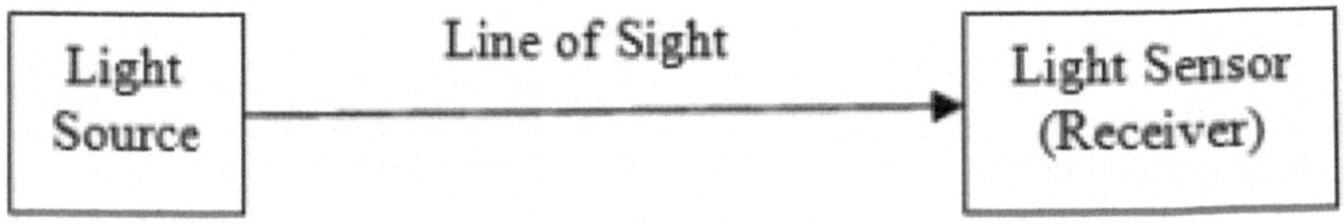

Fig. 1 Single-bit data transfer using Li-Fi

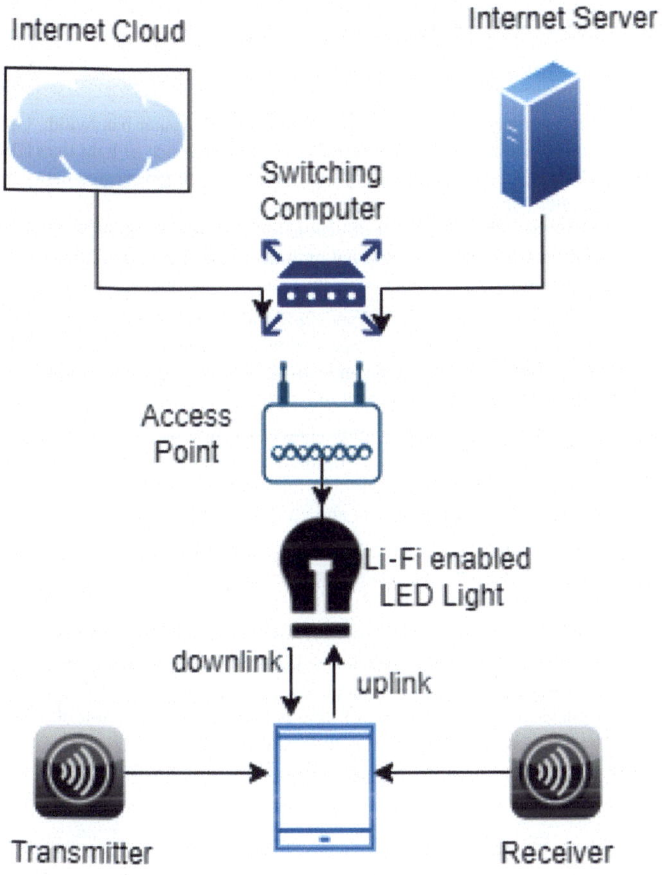

Fig. 2 Li-Fi-based data transfer model

The basic principle on which Li-Fi works is the transmission of digital 1 when the LED is on and the transition of digital 0 when the LED is off. A photodetector on the other end receives the data.

An array of LEDs can be used to send a vast amount of data. The rate of variation of LED intensity affects the speed of data transfer. The main factors to be considered while designing a Li-Fi-based system are a single light source, Good light source, and Illumination [22].

4 Characteristics of Li-Fi

Range	Li-Fi provides a free, easy-to-use, unlicensed greater bandwidth. Greater data density is provided by Li-Fi due to less interference of light than RF waves. The output speed is also very high because of this increased data density and greater bandwidth [20] Li-Fi extends the concept of visible light communication (VLC) to achieve fully networked, bi-directional multiuser communication (point-to-multipoint and multipoint-to-point), high speed and secure wireless communications
Efficiency	Because there are fewer components and no additional power input is required, pricing will be lower and thus cost-efficient. Everywhere LED bulbs are already present which can be used as the source of light to transfer data, therefore it is energy efficient too
Effects on the human body	Light does not affect the human body therefore no health hazards are caused by Li-Fi because light does not interfere with electronic circuitry [24], the technology remains safe and non-hazardous and not only it is safe for humans but safe for the environment and all living organisms
Anti-theft technology	Hacking and Data Leakage are quite difficult in the case of Li-Fi since light is confined by walls and opaque objects while other technologies use radio waves which can penetrate walls as atom size in the wall is smaller than the radio waves (Large waves), thus data might get breached easily
Data density	Density is the quantity per unit volume, unit length or unit area, and data density is the amount of data passing per unit volume. Li-Fi uses light to transfer data as the density of light is more so more amount of data can be transferred per second to the destination

4.1 Light Fidelity Versus Wireless Fidelity

Wi-Fi is deeply rooted in every facet of human society from households to basic government services such as transportation and ICT. Since we are deeply dependent on this mode of communication, we find it the most efficient technology for data transmission. Because of this, the concept of Li-Fi remains unfamiliar even though it has been around for years. Wi-Fi operates within the radio spectrum. An electromagnetic field is created when a radio frequency current is supplied to an antenna and it propagates through space enabling the users to access it through created access points. On the accessing device, there should be a wireless network adapter installed to connect to an access point. Since the visible light spectrum used by Li-Fi is 10,000 times larger than the entire radio spectrum [20], Li-Fi has access to a greater range of available frequencies.

Advantages of Li-Fi over Wi-Fi	Since light travels at a speed of around 3 × 108 m/s, it allows Li-Fi connections to transfer data very quickly resulting in faster data transmission and faster internet connections. Wi-Fi's one disadvantage among others is that it cannot penetrate through seawater as it easily gets absorbed [25, 26] and has more interference whereas Li-Fi can pass through salty seawater, can work in dense regions and has less interference
Range	The range of Wi-Fi is about 32 m (WLAN 802.11b), which varies with power and the type of antenna, whereas the range of Li-Fi is about 10 m [23]. The theoretical speed of Li-Fi is about 224 Gbps [20] which is much greater than Wi-Fi
Application	Wi-Fi is primarily used for browsing the internet and accessing the content available online on the web whereas Li-Fi has varied applications such as in hospitals, schools, undersea applications (divers health system, exploring undersea), offices, schools, indoor positioning systems and also used at home for browsing the internet and transferring the data and information, in indoor navigation system for blind people and in hazardous places where radio technology can be dangerous
Density of the environment	As Wi-Fi uses radio waves for data transmission, it is difficult for it to enter the denser environment and thus more data-dense areas affect the working of Wi-Fi whereas Li-Fi can work in high-density environments as it remains unaffected by data density
Interference	Wi-Fi network is susceptible to radio interference and thus the presence of several sources of radio waves can disrupt its working whereas Li-Fi does not have any interference issues similar to the radio waves [24]
Privacy	Wi-Fi is not secured as radio waves can penetrate walls and other objects whereas Li-Fi is highly secured as it uses light as a medium which can easily be stopped by opaque objects thus preventing unauthorized access to the Li-Fi connection From the data above, we can conclude that there are many advantages of Li-Fi over Wi-Fi but Li-Fi also have some disadvantage such as its range, light can easily be stopped by opaque objects hence it cannot cover large distances and limit its range to around 10 m [23] Thus, Li-Fi can act as a high-speed technology which can replace Wi-Fi in short range to improve the performance of existing technology and provide relief to a dense harmful spectrum of radio waves which is used in Wi-Fi connections

5 Research Paper Counts on Various Platforms

A survey has been done for the publication in the area of "Li-Fi" with popular publishers to see the research trends. The report is given in Table 2. This report says that Li-Fi technology is becoming more popular after 2017.

Table 2 Research paper counts in various platforms

Years	ScienceDirect	IEEE	ACM	Springer
2022		4	–	5
2021	4	43	–	15
2020	3	40	1	14
2019	3	56	–	11
2018	2	56	–	8
2017	1	41	–	5
2016	–	38	–	–
2015	1	36	1	–
2014	–	18	–	–
2013	–	1	–	–

6 Applications

6.1 *Underwater Communication*

Data is communicated using underwater Li-Fi by varying the amount of light that specialized underwater LED transmitters emit. In contrast to air, light spreads more quickly in water, improving signal speed. However, the range of effective communication is significantly hampered by the attenuation of light in water caused by absorption and scattering. Specific light wavelengths, such blue or green, are frequently chosen to increase communication range and reliability in order to address these problems. It is not possible to use Wi-Fi in underwater communication because, in water, the Radio Waves gradually start to die before reaching the destination. On the other hand, Li-Fi can be used because Li-Fi, which requires light as the medium of transmission of data, can penetrate deep water. Acoustic communication which is SONAR communication, The devices used in it are costly and have a low data rate of 10 Kbps [27] due to the slow speed of waves in water. So, an alternative method other than acoustic communication was required and thus, for high speed and better data transmission Li-Fi is used as an alternative.

6.2 *Vehicle-to-Vehicle Communication*

With time, the number of vehicles on the road is gradually increasing. Therefore, the number of accidents and deaths is increasing too. It can be reduced if the vehicles are made autonomous. This can be possible by using Li-Fi for connecting vehicles on the road. Since there is a headlight and backlight in a vehicle, the [28] paper proposed two scenarios when the first car is behind the other car, Fig. 3 and when the two

Fig. 3 Li-Fi for connecting vehicles in, the same direction

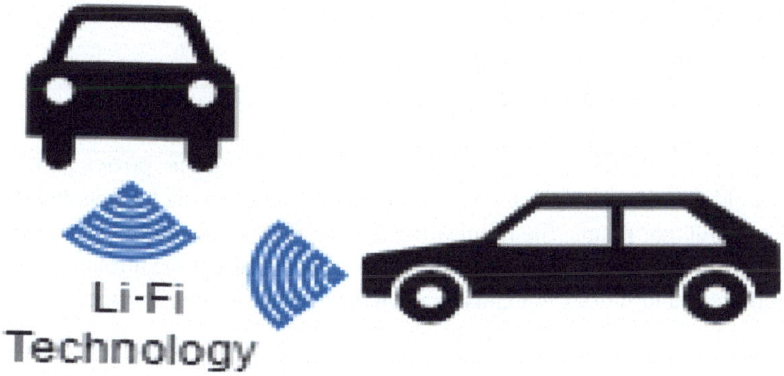

Fig. 4 Li-Fi for connecting vehicles, toward each other

cars are coming toward each other, Fig. 4. In both scenarios, the speed information will be transmitted to the cars and thus, messages like 'to slow down' will also be displayed.

LED is a valid light source for vehicle-to-vehicle communication [29]:

- Provides a more intense light beam.
- Provides a wide range of visibility.
- They have a long lifespan.
- No harm in LED lights.
- Cheaper.

Important features and captures of vehicle-to-vehicle communication:

- Speed of the vehicles around.
- Changing lane information.
- Information related to brakes.

6.3 Hospital

Wi-Fi transfers data through RF waves. Using Wi-Fi in hospitals, waves with radio frequency will affect the human body. So, it is better to use Li-Fi instead of using Wi-Fi. Li-Fi is used in Patient monitoring systems because of its speed and efficiency. The [30] proposed the working model of a patient monitoring system using Li-Fi, which requires many sensors. It also delivers the patient report to the appropriate individual without the need for human interaction. For transportation purposes, a robot was proposed in [31], which was named HOSPI. It has storage space, a security camera, weight of 120 kg with 20 kg of holding capacity.

6.4 Thermal Power Plants

As there are already too many radiations present in an atomic power plant, using Wi-Fi would not be the right choice. Thus, using Li-Fi in such areas would save time. As in Nuclear plants, the monitoring system must be faster, which is provided by Li-Fi. Also, data transmission through Li-Fi doesn't affect the human body. Li-Fi is cost-efficient and it does not cause any electromagnetic interference.

6.5 Schools

Li-Fi is faster and cheaper than other data-transmitting technologies. With the growing number of internet devices, it is getting very difficult to provide the internet speed which is required so installing Li-Fi in schools provides additional bandwidth and reduces network congestion, helping teachers and students to download or provide educational videos without any issue.

6.6 Li-Fi in Airways

Generally, we have to switch off our mobile phones in aircraft to prevent overlapping of signals associated with navigation used by aircraft. Since Li-Fi does not interfere with RF waves [24] it can be used for surfing the internet without any issue.

6.7 Li-Fi in the Indoor Navigation System for Blind People

Li-Fi is faster and more secure, therefore a navigation system for blind people is developed in the paper [32]. This system is like a roadmap for blind people for guidance in indoor places. When a person enters a room, the light coming from the lamp which is fixed on the ceiling is transmitted to the light receiver which is carried by the person. Then, the message is provided to the person through headphones by voice processing. For example, if the person enters the bedroom, the message is passed "You entered in the bedroom".

7 Various Other Types of Research

Author	Problem addressed	Outcome	Citations (as of April-2022)
Sharma et al. [28]	A small range of network bandwidth and greater latency delay	Framework which includes Li-Fi technology integrated with cloud computing	24
Kaur et al. [33]	It contains the study and research of various papers	Study and research of various papers	0
Margret Sharmila et al. [34]	Data loss in case of some emergency	Transfer of all the important data and information using high-speed Li-Fi technology in case of some emergency	0
Chaudhuri et al. [35]	A small range of frequencies of Wi-Fi	Most of the information, modulation techniques, etc. regarding Li-Fi technology	2
Singh et al. [36]	Limitations of radio wave technologies	Proposed a method to increase the range of Li-Fi data transfer	0
Durán et al. [37]	Study on QoS	Presents a detailed explanation of quality of service in Li-Fi technology	0
Fazea et al. [38]	Comparative study	Comparative study of 5G, Wi-Fi and Li-Fi. Discovered 5G in applications of the Internet of Things. A combination of other technologies such as Li-Fi with IoT is also mentioned	0
Durga et al. [39]	Health of the divers	A monitoring system is addressed in the paper which keeps the check on all the health parameters of divers	0

(continued)

(continued)

Author	Problem addressed	Outcome	Citations (as of April-2022)
Pacharne et al. [40]	Increase of road accidents	Proposed a system which transfers all the necessary information of a vehicle to the nearby vehicle using efficient Li-Fi technology	1
Swami et al. [32]	Increase of network traffic	Detailed explanation of Li-FI and Enhancement of Li-Fi for reducing traffic on cellular networks	1
Duraisamy et al. [41]	Study Paper	Splitting of light to transfer data from one source to many destinations. Understanding various characteristics of Li-Fi	0
Degala et al. [42]	Exponential increase of radio devices	Proposed a system of Li-Fi which will replace radio waves data transmitting technologies	0
Balaganesh et al. [43]	Future scope study paper	Analyzing the existing Li-Fi technology and exploring what enhancement can be done for more efficiency in future	0
Debnath et al. [44]	Study paper	Proposed a Li-Fi model for transferring data in a single direction to free space where the device will receive the information	0
Park et al. [45]	Objects in between the data transmission using Li-Fi	Proposed an algorithm and checked whether it was advantageous or not	0
Madjarov et al. [46]	It is difficult to find a way into a new building such as mall, hospital, etc.	A positioning system is introduced in this paper which gives direction to the people in building using Li-Fi technology	0
Gadgil et al. [47]	Corruption in ration shops	Proposed a system which will automate everything which was previously manual in a ration shop thus stopping the corruption	1
Mansingh et al. [48]	Slow data transmission speed at home	Proposed method which includes Li-Fi as the data transmission technology in home automation thus increasing the data transmission speed	4

(continued)

(continued)

Author	Problem addressed	Outcome	Citations (as of April-2022)
Sabareeswaran et al. [49]	Slow speed of Wi-Fi cannot hold all the users	Both hardware and software projects are introduced in this paper. Software project includes an android application which uses Li-Fi to transfer text messages. Hardware project transfers audio using LED, speaker and solar cell	0
Purohit et al. [50]	Exploring Li-Fi models	Proposed a model which downloads all the information in the form of text, photos and videos as soon as the user put their phones in the exhibit	3
Buttan et al. [51]	Study of Li-Fi in medical research	The impact of Li-Fi, on medical research developments impacting the human eye is examined in the following paper	0
Mukku et al. [52]	Li-FI used as a transceiver	Focuses on creating a custom Li-Fi transceiver for an Industry 4.0 learning factory that supports point-to-point and full-duplex connectivity	0
Swarnkar et al. [53]	Li-FI as a complementary technology to Wi-Fi	This paper discusses the evolution of Li-Fi, its design, requirements, and obstacles, as well as how existing protocols can be linked with it and its future scope	1
Singh et al. [54]	Increasing network traffic due to a large number of cell phones and other devicrs	The study offered Li-Fi as a way for reducing network traffic, as well as its application in business and innovation, and explained Li-Fi conclusions	0
Jaswal et al. [55]	Need for an positioning system for blind people	Proposed a model for blind people, an audio will be sent giving directions. This model is based on Li-Fi technology	0
Ibhaze et al. [56]	Study Paper	Explains the deployment of Li-Fi, presents its benefits and limitations. Future of IoT is also described in the paper	4
Agrawal et al. [57]	Study on vehicle data transmission	Enhancement of safe driving by passing important data and information to the vehicle through visible light communication	0
Cha et al. [58]	Scattered data and information	Study on Datalake framework using Li-Fi	0

(continued)

(continued)

Author	Problem addressed	Outcome	Citations (as of April-2022)
Panda et al. [59]	Less range of radio spectrum	Proposed a research to calculate the coordinates of an object and derived a mathematical form by using Li-Fi	1
Vijayalakshmi et al. [60]	Increasing radio traffic	Li-Fi to transfer data while traveling in train	1
Riurean et al. [61]	Interference of radio wave data transmission with radio devices at hospital	Wireless medical system is addressed in the paper using both Li-Fi and radio wave technologies	4
Sandeep et al. [62]	Wi-Fi is slower	Proposed a model in which image gets transferred with the help of rasberry pi which converts image to bits and bits to image using light as the source	0
Kumawat et al. [63]	Limited bandwidth of radio wave technologies	Proposed the idea of using Li-Fi instead of Wi-Fi in metro, due to reasons such as cost effective, high speed, etc	0
Nor et al. [64]	Requirement of low latency and high speed network	Introduced integration of mmwave technologies and Li-Fi	8
Gupta et al. [65]	Health hazards by radio waves	Proposed efficient approach for secured data transfers using Li-Fi by integrating it to Arduino UNO	0
Lambrechts et al. [66]	Large investment required by radio wave technologies in rural areas	This paper contains applications, principles of visible light communication, benefits and limitations of Li-Fi	1
Paul et al. [67]	Study Paper	This paper has presented most of the information related to Li-Fi such as its application, security, characteristics and other parameters	0
Parthasaradi et al. [68]	Scarce spectrum of radio waves	Proposed a robotic vehicle which uses Li-Fi technology where light energy is collected from solar panels	1
Bao et al. [69]	Crisis of radio wave networks	Surveyed various data transmitting technologies with Li-Fi and mentioned challenges and its future scope	6

7.1 Limitations and Advantages

Li-Fi being a next-generation wireless communication technology has various advantages with some limitations. We'll discuss speed, security, range of Li-Fi, cost of installing, effect on electromagnetic waves, restraint mobility and interference to Li-Fi.

Speed	With time number of users and devices is increasing, and the demand for high speed internet is increasing too, so to fulfill this need of high speed data transmission we use Li-Fi. Around 1 Gbps of data rates has been reported in paper [70] and Li-Fi can be 150 times faster than Wi-Fi becoming one of the fastest ways of transmitting data
Security	Since Li-Fi uses light to transfer data and light does not penetrate opaque objects that means there is no risk of getting data leaked. This property of not penetrating opaque objects also comes with its limitations as objects can disturb the data transmission from sender to receiver
Effects on electromagnetic waves	Visible light does not affect electromagnetic waves, therefore Li-Fi too not affect electromagnetic waves because of this property, Li-Fi is differentiated from the other technologies as it is less dangerous compared to other RF wave technologies. It can be used in hazardous areas like thermal power plants, petrol stations, etc
Cost	Cost is also a major factor, the installation cost of Li-Fi is a little higher than the other technologies but in the long run, Li-Fi saves more money compared to other technologies. There are fewer components and no additional power input is required, LED is easily available and is cheaper too thus making Li-Fi cost-efficient
Range of Li-Fi	As light cannot penetrate opaque objects its range is less than around 10 m [23] with certain conditions like no other light source interfering with the current data transmitting light source and sunlight affecting the data transmission, it cannot be placed in sunlight. These limitations have been solved to a certain extent

8 Results and Discussion

The speed of Li-Fi technology is highest and comparable with the 5G technology. Since light travels at a speed of around 3 × 108 m/s, it allows Li-Fi connections to transfer data very quickly resulting in faster data transmission and faster internet connections while Wi-Fi's one disadvantage among others is that it cannot penetrate through seawater. So, Li-Fi is an interesting area of research and it is proved by the research work published in the major reputed publications and it is increasing year by year. It is more suitable for underwater communication as in comparison to air, light spreads more quickly in water, improving signal speed.

9 Conclusion

This chapter covers an overview of Li-Fi, its properties, and a comparison of various data transfer technologies such as Wi-Fi and Li-Fi, as well as applications such as underwater data transmission, its use in thermal power plants, hotels, hospitals, and shopping malls, drawbacks and advantages over competing technologies is also mentioned. Li-Fi is fast and secure, better than many technologies available but still, it is not used by people because of low awareness of this technology, it is difficult to switch from one technology to another. Wi-Fi has been used by people for a long time that's why people are not willing to switch to other technology. This chapter will raise awareness toward Li-Fi and might help people to switch to the new generation technology that is Li-Fi, but due to some of its limitations other technologies cannot be replaced by Li-Fi but it can act as a complementary technology to reduce the burden on others.

References

1. Tao W, Zhao L, Wang G, Liang R (2021) Review of the internet of things communication technologies in smart agriculture and challenges. Comput Electron Agric 189:106352. https://doi.org/10.1016/j.compag.2021.106352
2. Kaushal H, Kaddoum G (2016) Underwater optical wireless communication. IEEE Access 4:1518–1547
3. Rajagopal S, Roberts RD, Lim SK (2012) IEEE 802.15. 7 visible light communication: modulation schemes and dimming support. IEEE Commun Mag 50(3):72–82
4. Haas H (2011) Wireless data from every light bulb, TED: ideas worth spreading
5. Haas H, Yin L, Wang Y, Chen C (2015) What is lifi? J Lightwave Technol 34(6):1533–1544
6. Vainio JT (2000) Bluetooth security. In: Proceedings of Helsinki University of technology, telecommunications software and multimedia laboratory, seminar on internetworking: Ad Hoc networking. Springer
7. https://standards.ieee.org/ieee/802.11/5536/
8. Termizi SAA, Rahmat S, Qadeer MA (2016) LTE-U over Wi-Fi. In: 2016 Thirteenth international conference on wireless and optical communications networks (WOCN). IEEE, pp 1–7
9. Kamali B (2018) AeroMACS (An IEEE 802.16 standard-based technology for the next generation of air transportation systems) || The IEEE 802.16 standards and the WiMAX technology, pp 189–258. https://doi.org/10.1002/9781119281139
10. Safaric, Stanislav; Malaric, Kresimir (2006). IEEE ELMAR 2006, Zadar, Croatia 6 June –10 June 2006 proceedings ELMAR 2006—ZigBee wireless standard, pp 259–262. https://doi.org/10.1109/elmar.2006.329562
11. Rao YS, Yeung W-C, Kripalani A (2000) Third-generation (3G) radio access standards, In: IEEE 16th INTERNATIONAL CONFERENCE ON COMMUNICATION TECHNOlogy (ICCT'00), Beijing, China 21–25 Aug 2000, WCC 2000—ICCT 2000. 2000 international conference on communication technology proceedings (Cat. No.00EX420), vol 2, pp 1017–1023. https://doi.org/10.1109/ICCT.2000.890849
12. Kharga P (2015) Gi-Fi technology: a technology with standard features. Int J Comput Appl 4–6
13. Bhandarkar S, Kammar, R (2016) 4G Technology. https://doi.org/10.5281/zenodo.158935

14. Pocero L, Tsampas S, Mylonas G, Amaxilatis D (2020). A comparative study of LoRa and IEEE 802.15.4-based IoT deployments inside school buildings. IEEE Access 1–1. https://doi.org/10.1109/ACCESS.2020.3020685
15. Gallagher JC, DeVine ME (2019) Fifth-generation (5G) telecommunications technologies: issues for congress. Congressional Res Ser 1(30):1–39
16. Haas H, Chen C, O'Brien D (2017) A guide to wireless networking by light. Prog Quantum Electron 55:88–111
17. Raza U, Kulkarni P, Sooriyabandara M (2017) Low power wide area networks: an overview. IEEE Commun Surv Tutorials 19(2):855–873. https://doi.org/10.1109/COMST.2017.2652320
18. Altunay Ö (2020) Host-to-host tcp/ip connection over serial ports using li-fi for internet connection. Master's thesis, Fen Bilimleri Enstitüsü
19. Ayyash M, Elgala H, Khreishah A, Jungnickel V, Little T, Shao S, Freund R (2016) Coexistence of WiFi and LiFi toward 5G: concepts, opportunities, and challenges. IEEE Commun Mag 54(2):64–71
20. https://news.mit.edu/2018/wireless-communication-through-water-air-0822
21. Zeng L, O'Brien D, Minh H; Faulkner G, Lee K, Daekwang Jung, D; Oh YJ, Won ET (2009) High data rate multiple input multiple output (MIMO) optical wireless communications using white led lighting. 27(9):1654–1662. https://doi.org/10.1109/jsac.2009.091215
22. Pathak PH, Feng X, Hu P, Mohapatra P (2015) Visible light communication, networking, and sensing: a survey, potential and challenges. IEEE Commun Surv Tutorials 17(4):2047–2077
23. Kulkarni S, Darekar A, Joshi P (2016) A survey on Li-Fi technology. In: 2016 International conference on wireless communications, signal processing and networking (WiSPNET). IEEE, pp. 1624–1625
24. Petrosino A, Striccoli D, Romanov O, Boggia G, Grieco LA (2023) Light fidelity for Internet of Things: a survey. Opt Switching Netw 48:100732. ISSN 1573-4277. https://doi.org/10.1016/j.osn.2023.100732
25. https://lifi.co/lifi-misconceptions/
26. Kulkarni S, Darekar A, Shirol S (2017) Proposed framework for V2V communication using Li-Fi technology. In: IEEE 2017 international conference on circuits, controls, and communications (CCUBE), Bangalore, India, 15 Dec 2017 16 Dec 2017, 2017 International conference on circuits, controls, and communications (CCUBE), pp 187–190
27. Abdulsalam, Noof Al; Hajri, Raya Al; Abri, Zahra Al; Lawati, Zainab Al; Bait-Suwailam, Mohammed M (2015) Design and implementation of a vehicle to vehicle communication system using Li-Fi technology. IEEE 2015 International conference on information and communication technology research (ICTRC), Abu Dhabi, United Arab Emirates 17 May 2015–19-May 2015, International conference on information and communication technology research (ICTRC), pp 136–139
28. Sharma PK, Ryu JH, Park KY et al (2018) Li-Fi based on security cloud framework for future IT environment. Hum Cent Comput Inf Sci 8:23
29. Sudha S, Indumathy D, Lavanya A, Nishanthi M, Sheeba D Merline; Anand V (2016) Patient monitoring in the hospital management using Li-Fi. In: IEEE 2016 IEEE technological innovations in ICT for agriculture and rural development (TIAR), Chennai, India 15 July 2016 to 16 July 2016, 2016 IEEE Technological Innovations in ICT for Agriculture and Rural Development (TIAR), pp 93–96
30. Murai, R, Sakai T, Kawano H, Matsukawa Y, Honda Y, Campbell K (2012) A novel visible light communication system for enhanced control of autonomous delivery robots in a hospital. In: Proceedings of the IEEE/SICE international symposium on system integration (SII), pp 510–516
31. Kalid AM, Cossu G, Corsini R, Choudhury P, Ciaramella E (2012) 1-Gb/s transmission over a phosphorescent while led by using rate-adaptive discrete Multitone Modulation. IEEE Photonics J 4(5):1465–1473
32. Swami NV, Sirsat NB, Holambe PR (2016) Light fidelity (Li-Fi): in mobile communication and ubiquitous computing applications. In: Chakrabarti A, Sharma N, Balas V (eds) Advances in computing applications. Springer, Singapore

33. Kaur H (2021) Li-Fi technology—a study on a modernistic way of communication. In: Marriwala N, Tripathi CC, Kumar D, Jain S (eds) Mobile radio communications and 5G networks. Lecture notes in networks and systems, vol 140. Springer, Singapore
34. Margret Sharmila F, Akshaya M, Aksshaya DR, Harini KV, Harshini S (2019) Li-Fi based building safety system. In: Hemanth J, Fernando X, Lafata P, Baig Z (eds) International conference on intelligent data communication technologies and Internet of Things (ICICI) 2018, ICICI 2018. Lecture notes on data engineering and communications technologies, vol 26. Springer, Cham
35. Chaudhuri RR, Dutta K, Saha A (2015) Review on Li-Fi technology. In: Lakshminarayanan V, Bhattacharya I (eds) Advances in optical science and engineering. In: Springer proceedings in physics, vol 166. Springer, New Delhi
36. Singh S, Choudhury T, Joshi P, Chhokar RL (2018) A modified technique for Li-Fi range extension. In: Bhateja V, Tavares J, Rani B, Prasad V, Raju K (eds) Proceedings of the second international conference on computational intelligence and informatics. Advances in intelligent systems and computing, vol 712. Springer, Singapore
37. Durán JMP, Parra OJS, Sarmiento DAL (2019). Quality of service applied to Li-Fi networks in 5th generation environments. In: Renault É, Boumerdassi S, Bouzefrane S (eds) Mobile, secure, and programmable networking. MSPN 2018. Lecture notes in computer science, vol 11005
38. Fazea Y, Mohammed F, Al-Nahari A (2022) A review on 5G technology in IoT-application based on light fidelity (Li-Fi) indoor communication. In: Saeed F, Mohammed F, Ghaleb F (eds) Advances on intelligent informatics and computing. IRICT 2021. Lecture notes on data engineering and communications technologies, vol 127
39. Durga R, Venkatesh R, Selvaraj D (2020) Underwater Li-Fi communication for monitoring the divers health. In: Hemanth D, Kumar V, Malathi S, Castillo O, Patrut B (eds) Emerging trends in computing and expert technology. COMET 2019. Lecture notes on data engineering and communications technologies, vol 35
40. Pacharne S, Kulkarni V (2021) Vehicle-to-Vehicle driver safety-related data transmission and reception using Li-Fi technology. In: Merchant SN, Warhade K, Adhikari D (eds) Advances in signal and data processing. Lecture notes in electrical engineering, vol 703. Springer, Singapore
41. Duraisamy B (2021) Multipoint data transmission using Li-Fi with LGS formation. In: Peng SL, Hsieh SY, Gopalakrishnan S, Duraisamy B (eds) Intelligent computing and innovation on data science. Lecture notes in networks and systems, vol 248. Springer, Singapore
42. Degala S, Degala SS (2018) A three level architecture for wireless communication using Li-Fi. In: Meesad P, Sodsee S, Unger H (eds) Recent advances in information and communication technology 2017. IC2IT 2017. Advances in intelligent systems and computing, vol 566. Springer, Cham
43. Balaganesh D (2020) Inadequacy of Li-Fi disentangles by Laser, Polarizing Beam, Solar, and formation. In: Peng SL, Son LH, Suseendran G, Balaganesh D (eds) Intelligent computing and innovation on data science. Lecture notes in networks and systems, vol 118. Springer, Singapore
44. Debnath S, Bhowmik BB (2021) Design of a low-cost Li-Fi system using table lamp. In: Mandal J, Mukhopadhyay S, Roy A (eds) Applications of Internet of Things. Lecture notes in networks and systems, vol 137. Springer, Singapore
45. Park H (2021) Multiuser channel access with RTS-CTS algorithm for Li-Fi communications. In: Barolli L, Poniszewska-Maranda A, Park H (eds) Innovative mobile and internet services in ubiquitous computing, IMIS 2020. Advances in intelligent systems and computing, vol 1195. Springer, Cham
46. Madjarov I, Damoiseaux JL, Iguernaissi R (2021) Towards designing a Li-Fi-based indoor positioning and navigation system in an IoT context. In: Arai K (eds) Advances in information and communication, FICC 2021. Advances in intelligent systems and computing, vol 1363. Springer, Cham
47. Gadgil AA, Bendigeri AS, Halagi DD, Prakash G, Nadgouda SA (2018) Secure ration dispensing system using HAN and geofencing through Li-Fi. In: Pawar P, Ronge B,

Balasubramaniam R, Seshabhattar S (eds) Techno-Societal 2016. ICATSA 2016. Springer, Cham
48. Mansingh PMB, Yuvaraju M (2021) Improved data transmission using Li-Fi technology for home automation application. J Ambient Intell Human Comput 12:5581–5588
49. Sabareeswaran S, Madumitha G, Shruthi M (2020) Text and audio transfer using LI-FI technology. In: Kumar L, Jayashree L, Manimegalai R (eds) Proceedings of international conference on artificial intelligence, smart grid and smart city applications. AISGSC 2019 2019. Springer, Cham
50. Purohit S, Tolani U, Goswami S (2018) OnLight information broadcasting in exhibitions using Li-Fi. In: Satapathy S, Bhateja V, Chowdary P, Chakravarthy V, Anguera J (eds) Proceedings of 2nd international conference on micro-electronics, electromagnetics and telecommunications. Lecture notes in electrical engineering, vol 434. Springer, Singapore
51. Buttan Y, Saxena K (2021) Creating a biological intranet with the help of medical sciences and Li-Fi. In: Goyal D, Chaturvedi P, Nagar AK, Purohit S (eds) Proceedings of second international conference on smart energy and communication. Algorithms for intelligent systems. Springer, Singapore
52. Mukku VD, Lang S, Reggelin T, Reichardt P (2021). Design of a Li-Fi transceiver for distributed factory planning applications. In: Dolgui A, Bernard A, Lemoine D, von Cieminski G, Romero D (eds) Advances in production management systems. Artificial intelligence for sustainable and resilient production systems, APMS 2021. IFIP advances in information and communication technology, vol 630. Springer, Cham
53. Swarnkar M, Bhadoria RS, Arya KV (2018) Architectural building protocols for Li-Fi (Light Fidelity). In: Arya K, Bhadoria R, Chaudhari N (eds) Emerging wireless communication and network technologies. Springer, Singapore
54. Singh D (2022) Li-Fi: an emerging technology. In: Senjyu T, Mahalle PN, Perumal T, Joshi A (eds) ICT with intelligent applications. Smart innovation, systems and technologies, vol 248. Springer, Singapore
55. Jaswal A, Mahal V, Ahuja B, Khatri A (2020) Navigation system for visually impaired using Li-Fi technology. In: Pant M, Sharma T, Verma O, Singla R, Sikander A (eds) Soft computing: theories and applications. Advances in intelligent systems and computing, vol 1053. Springer, Singapore
56. Ibhaze AE, Orukpe PE, Edeko FO (2020) Li-Fi prospect in Internet of Things network. In: Arai K, Kapoor S, Bhatia R (eds) Advances in information and communication, FICC 2020. Advances in intelligent systems and computing, vol 1129. Springer, Cham
57. Agrawal R, Thangadurai N (2020) Future of smart vehicle communication by using Li-Fi technology. In: Kadambi G, Kumar P, Palade V (eds) Emerging trends in photonics, signal processing and communication engineering. Lecture notes in electrical engineering, vol 649. Springer, Singapore
58. Cha B, Park S, Shin BC, Kim J (2020) Draft design of Li-Fi based acquisition layer of Datalake framework for IIoT and smart factory. In: Barolli L, Nishino H, Enokido T, Takizawa M (eds) Advances in networked-based information systems, NBiS—2019. Advances in intelligent systems and computing, vol 1036. Springer, Cham
59. Panda B, Beria H, Pradhan C (2021) Deployment of Li-Fi in indoor positioning systems. Int. j. inf. tecnol. 13:123–130
60. Vijayalakshmi BA, Nesasudha M (2020) Li-Fi delivering broad band and multimedia content under dimming in train. Opt Quant Electron 52:518
61. Riurean S, Antipova T, Rocha A, Leba M, Ionica A (2019) Li-Fi embedded wireless integrated medical assistance system. In: Rocha Á, Adeli H, Reis L, Costanzo S (eds) New knowledge in information systems and technologies. WorldCIST'19 2019. Advances in intelligent systems and computing, vol 931. Springer, Cham
62. Sandeep SC, Sreenivasa Reddy P, Shrenik M, Farheen S, Praveen Kumar D (2021) Utilizing a raspberry Pi for transmitting image using Li-Fi transceiver. In: Ranganathan G, Chen J, Rocha Á (eds) Inventive communication and computational technologies. Lecture notes in networks and systems, vol 145. Springer, Singapore

63. Kumawat A, Sriram Kumar D (2020) Light-Fidelity (Li-Fi), a new approach for internet inside running metro. In: Janyani V, Singh G, Tiwari M, Ismail T (eds) Optical and wireless technologies. Lecture notes in electrical engineering, vol 648. Springer, Singapore
64. Nor AM, Mohamed EM (2019) Li-Fi positioning for efficient millimeter wave beamforming training in indoor environment. Mobile Netw Appl 24:517–531
65. Gupta S, Sarkar M, Kaur H, Agrebi M, Roy A (2022) An efficient data transferring through Li-Fi technology: a smart home appliance. In: Kumar R, Sharma R, Pattnaik PK (eds) Multimedia technologies in the Internet of Things environment, vol 3. Studies in big data, vol 108. Springer, Singapore
66. Lambrechts W, Sinha S (2019) A theoretical analysis of Li-Fi: a last mile solution. In: Last mile internet access for emerging economies. Lecture notes in networks and systems, vol 77. Springer, Cham
67. Paul R, Sebastian N, Yadukrishnan PS, Vinod P (2021) Study on data transmission using Li-Fi in vehicle to vehicle anti-collision system. In: Pandian A, Fernando X, Islam SMS (eds) Computer networks, big data and IoT. Lecture notes on data engineering and communications technologies, vol 66. Springer, Singapore
68. Parthasaradi V, Kailasapathi P (2019) A novel MFCC-NN learning model for voice communication through Li-Fi for motion control of a robotic vehicle. Soft Comput 23:8651–8660
69. Bao X, Yu G, Dai J et al (2015) Li-Fi: Light fidelity-a survey. Wirel Netw 21:1879–1889
70. Botre MR, Askhedkar, AR (2019) LiFi and voice based indoor navigation system for visually impaired people. In: IEEE 2019 IEEE Pune section international conference (PuneCon), Pune, India 18 Dec 2019–20 Dec 2019, 2019 IEEE Pune Section International Conference (PuneCon), pp 1–4

Securing the Future: The Intersection of IoT, Cybersecurity Implementation, and AI Management

Vikas Kumar, Sangeeta Rani, Tanvi Rustagi, and Ajay Kumar

Abstract The IoT (Internet of Things) is a type of web which connects one device with other with the use of web. The IoT guide to send data among machines, recording and looking at devices and other input. It makes devices smart by allowing them to send data and removing manual tasks, without the use of any physical appearance. Cyber-security is a way that helps organizations to safeguard their devices, data, functions, area, and devices from cyber-attacks and unauthorized access. Cyber-security threats are growing faster in large as attackers use new method and common engineering to get ransom from companies and individuals, disturb business functions and takeaway or destroy important info. In this we are going to study the impact of IoT on Cyber Security and how we can deal with this. Although IoT devices have large no of uses and there are various impacts of these. It is very much possible to prevent ourselves from getting impacted from these, will be discussing the various method to prevent them. To save from these activities, the company needs technology cyber-security methods and effective process to find and stop threats and remove a cyber-security breach. Technology is important in protecting companies, devices, networks, and machine. Important cyber-security technologies are antivirus software, next-generation firewalls (NGFWs) and email related solutions. It's relevant to keep in mind that your technical profile is only as good as the frequency and quantity. This chapter aims to identify and characterize the various categories of IoT applications, assess potential security risks related to them and their effects on these applications, and present a linear programming methodology (LPP) for providing funding for multiple IoT and vulnerability projects. To find IoT application systems and associated cyber vulnerabilities, a comprehensive investigation and assessment of the IoT Cyber Risk Management framework has been conducted.

V. Kumar · S. Rani (✉) · T. Rustagi
Department of Computer Science and Engineering, World College of Technology and Management, Gurgaon, Haryana, India
e-mail: sangeeta.sept@gmail.com

A. Kumar (✉)
Department of Mechanical Engineering, School of Engineering and Technology, JECRC University, Jaipur, Rajasthan, India
e-mail: ajay.kumar30886@gmail.com

A. Kumar et al. (eds.), *Trends in Mechatronics Systems*, Emerging Trends in Mechatronics, https://doi.org/10.1007/978-981-97-9108-8_7

Keywords Internet of Things · Cyber security · Risk management · Linear programming · IT assets · Cyber-attack · Risk calculations

1 Introduction

Due to the IoT, or the online world of Things which has created devices that can communicate and work together, businesses are now developing new processes. IoT, cyber-security assault is widespread and getting very common, causing. Different types of issue for user and organizations in the way of conformity image, money, in company operations. The extraordinary profile of IoT products in areas like smart utilities, surveillance of the environment, healthcare mechanisms, intelligent production, and transportation is accountable for the steep rise in cyber-attacks. The security of the IoT is a problem, because of the fluctuating and brief nature of connections between devices, the variety of forces that can operate inside IoT systems, and resource restrictions [1–5].

Despite inadequate security measures, current threat evaluation approaches are unsuitable for fluid systems like the IoT. In this respect, current risk evaluation methods are not suitable for the assessment of cyber-security risk because to the complex nature of healthcare IoT systems, which exposes multiple attack avenues to adversaries. Organizations may be able to establish IoT security measures without unintentionally increasing cyber threats by constructing IoT devices on an open-source framework [6–8]. By safeguarding and protecting one's data, IoT cyber-security target weaker the threat of cyber-security for both companies and customers. IoT cyber-security potential as well as issues as new safeguards continue to develop. Older studies tended to focus on the technical elements of IoT security. This study designs a four-layer IoT cyber security control system, reviews the literature on the gap in IoT cyber-security risk control is addressed by IoT security technology and cyber risk control frameworks.

The cyber-security of IoT is mostly used by countries and education system to use methods and laws to get a high degree of cyber-security. The United States, China, and the United Kingdom are the three parts of the world affected by IoT cyber-security threats, especially by smart home attacks. China's Cyber-security Law (CSL) was initiated on June 1, 2017. The Cyberspace Administration of China (CAC) is the primary governmental authority to supervise and enforce the CSL. The CSL regulates cyber-security from different aspects, coordinating tracking, alerting, and rescue efforts on the Chinese mainland, as well as operating networks and data security. Europe has advanced in several fields, such as energy, vehicles, and residential, in cyber-security. Figure 1 shows security issues in various home based IoT devices.

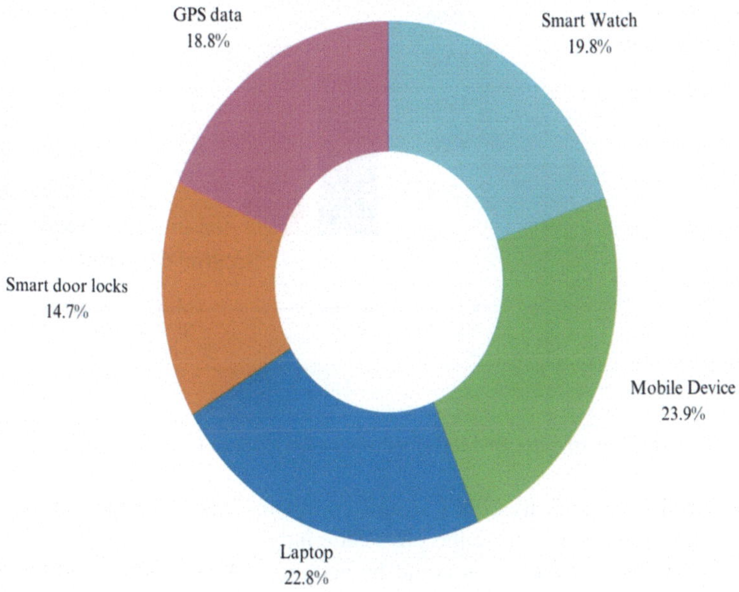

Fig. 1 Home-based IoT devices in cyber security issue from year 2018 to 2022

2 Preliminaries

This section defines the IoT problem in day-to-day life and provides a general overview of cyber security management related to IoT. Given the unique characteristics of cyber hazards and the flaws in IoT systems, understanding both IoT online-security innovations and current online risk prevention frameworks is important for creating a new IoT online risk mitigation system.

3 Notation of Problems and Definition

Security is a significant issue that needs to be resolved in order to advance IoT into the world of reality. The conventional systems of information (IS) security objectives (SO), which include privacy, reliability, as well as information accessibility [9], are aligned with security problems related to IoT. Additionally, there are additional safety concerns that seem to be IoT-generic. For instance, the combination of cloud-based computing and the Internet of Things opens IoT systems to web-induced flaws like those found in the top 5 list by Open Web Application Security Project (OWASP). Figure 2 highlights risk caused by the cloud that were discovered in 2015, according to the common susceptibility database [10]. As IoT solutions and services develop, the flaws that are present in cloud-based software will continue to leave an impact. These could pose a challenge to the ability of IoT solutions to survive. Items that

Fig. 2 No of security challenges relating to IoT

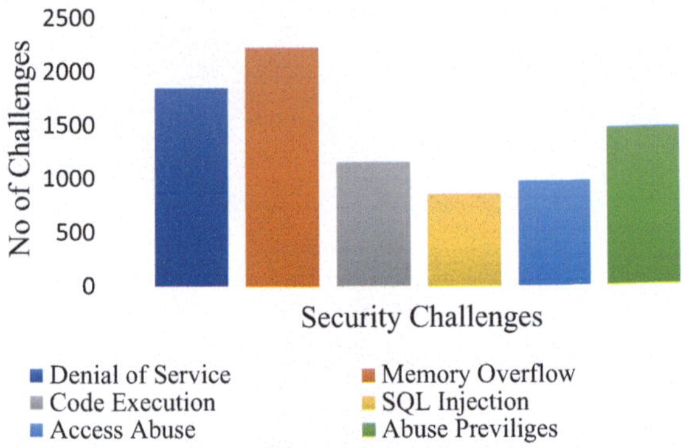

are poorly made, out-of-date or fake pose very real threats to IoT-enabled structures [11].

Additionally, the majority of IoT security problems arise from the system's own inbuilt weaknesses, which leave the physical configuration open to several attacks [12]. The sources of information could be network cables or ports and firmware, which is electronics (device), programmed system, data, or hardware itself. Additionally, the system is susceptible to network defined threats and protocol failure for the reason to the dual-direction interactions between objects [13]. This normally spread the threat to area by different cyber-security expert. Security experts are mindful of this and take measures to handle the safety risks that arise. Protecting the Internet of Things and the internet connections through which they are linked is also essential [14]. Production equipment, energy systems, building automated processes, and other IoT personal technology that employees bring to their jobs are examples of IoT gadgets [15] that are used in commercial operations.

3.1 *Cyber-Security in IoT Architecture*

To Over Come all these issues cyber security Architecture has been categorized in different layers. Security measures should be considered for the whole structure because each part of the Internet of Things design has a different security issue and interacts with other tiers. [16]. Figure 3 shows layer-wise Cyber Security architecture.

Cyber-security method by the eye for the design let us a proper interrogation sight of the IoT cyber method. The following focuses on layer-level cyber-security issues and resolutions and is based on Lee's five-layer corporate IoT architecture [17].

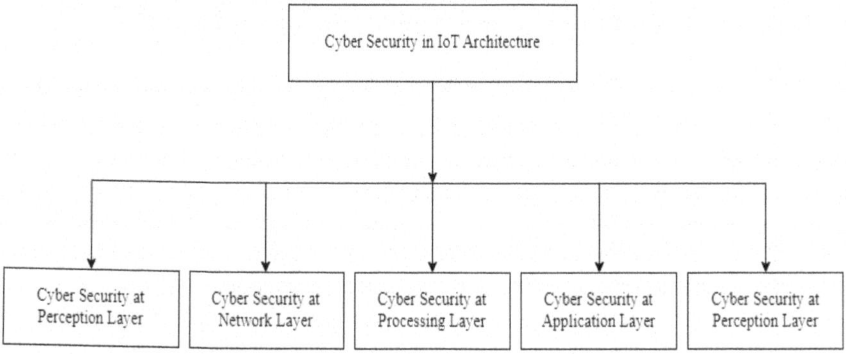

Fig. 3 Layer-wise cyber security architecture

3.2 *Cyber-Security Risk Management*

Authentication, encryption, access control, data protection, identity management, and intrusion detection and prevention are the parts of the Security Architecture for IoT [18]. Together, these guard networks and IoT devices from possible online dangers. The cyber hazards that a company face may be measured using both qualitative and quantitative methods [19]. Although it lacks a model to adhere to, the National Institute of Standards and Technology (NIST) framework is widely documented and offers assistance on risk assessment and management implementation. NIST evaluates hazards qualitatively, as shown in Fig. 4, and does not have an IoT effect assessment methodology [20].

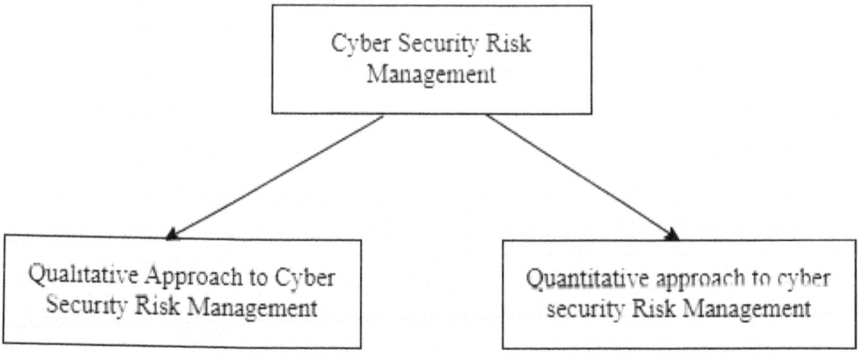

Fig. 4 Types of cyber security in IoT architecture based on risk management

4 Layered Framework of IoT Cyber Risk Management

For the benefit of IoT protection dealers are working in creating a cost-preventive cyber-security risk management strategy, the four-tier control method is provided as shown in Fig. 5. To organize and analyze risk management actions tier by tier lacking missing focus of the overall security measures challenges, suggested structure identifies the key elements affecting risks related to cyber-security and organizes those elements into the four layers. Ecosystem layer, infra structures tier, risk evaluation, and performance tiers make up the risk assessment methodology [21].

4.1 Ecosystem Layer

It tells us about people dealing with IoT devices cooperatively and aggressively. Developers of IoT cyber-security technologies, external users and consumers, enemies, administrations, and standardization organizations belong to the ecosystem's major competitors [22]. Cyber-security managers must react quickly in reaction to change in the IoT cyber environment to establish suitable safeguards in order to protect IoT systems.

4.2 Infrastructure Layer

Companies do assess their present IoT security architecture from both a technical and administrative viewpoint to develop a prospective cyber risk prevention plan.

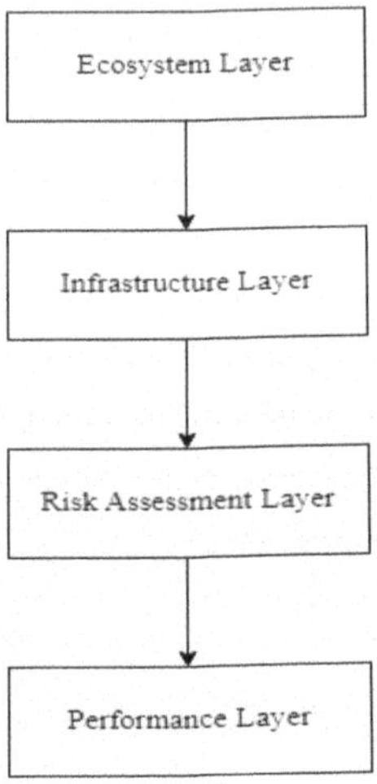

Fig. 5 The four-layer risk management framework

Nevertheless, the absence of organizational [23] and user support for protection is just as harmful to it as the absence of security solutions. These employees/internal consumers, organization, and security systems make up the IoT digital infrastructure layer, which supports an integrated strategy to cyber risk management [24].

4.3 Risk Assessment Layer

After knowing security structures, the organization must evaluate these present and conceivable future cyber hazards. The three main tasks that make IoT cyber risk assessment layer are recognizing risks, risk measurement, and allocation of resources.

The method of targeting attacks every time is an important component of the source providing model because, as mentioned before, cyber-attacks follow chances distributions (such as the Poisson probability method). A commensurate investment will help the defender realize their objective of raising defense likelihood. Defense [25] chances are binomial, meaning they can either be successful or unsuccessful. The overall money loss of IoT help-risks-assaults type is determined by the likelihood of each IoT help-risks-threat type and the occasion of penetrating attack combined. Chances of a fired attack (i.e., no successful defense) are computed as follows: x * (1 q), where n tells the nod of attacks per time and p is likelihood of a successful defense; x * (1 q) * y is all debt of money; and m tells the amount of money lost for each fired attack. Consider the IoT asset-risk-threat-1 as an example and considering that the present defense chance is roughly 0.8. Considering that the IoT help-risk-threat-1 has recently received 4000 threats per unit, each fired at a cost of Rs 500. The total amount of money lost because of the IoT help-risk-threat-1 is therefore close to Rs. 400,000 (4000 * (1 0.8) * (Rs. 1000). The most effective distribution of scarce resources across competing IoT asset-vulnerabilities-threats [26] is one of the issues in managing cyber-security. To decide how to allocate resources to address different IoT asset-risks-threats, a cost-effectiveness model is used. Corporate assets are translated into monetary units and comprise labor, IoT resources, and capital [27]. In order to efficiently distribute monetary funds for every IoT asset-risks-threat, this article employs a linear programming [28] approach It should be highlighted that no asset planning techniques have been used to make funding choices in security measures. For many business problems, including estimating, logistics planning, and manufacturing organizing, linear programming is an understandable and logical statistical modeling method that gives best method within few financial and operational constraints, and profile management. Since numerous cyber-security projects and the limited funds allocated for their management, the linear programming method can be helpful choices.

Figure 6 tells that whatever money is allocated will change the defense probabilities from the initial 0.8 defense probabilities. The defense chances are indicated on the x-axis from 0.8 to 0.88. Due to the increased likelihood of defense, the money debt from the IoT help-risks-threat-1 will be smaller the more financial resources are allocated. The final cost is the addition of the money lost from fired attack for given cyber investing [29] cost of aimed define chances. In the, e.g., line graphs are considered for dual the money loss and the investing cost with the slot of 0.8 and 0.88. A rapid growth of the online investing rate is considered from a defense chance of 0.88 and higher because of constant growth in difference rate of technology. In this e.g., the method for the money debt is ML = Rs. 400,000–Rs. 4,000,000 * (a – 0.8), where a is the defense probability. The method to find investing cost is IC = Rs. 2,000,000 * (a – 0.8). The method of final sum is the total of ML and IC: FS = FL + IC = Rs. 400,000–Rs. 1,000,000 * (a – 0.8). Figure 7 presented the complete cost benefit analysis.

Fig. 6 Example risk matrix

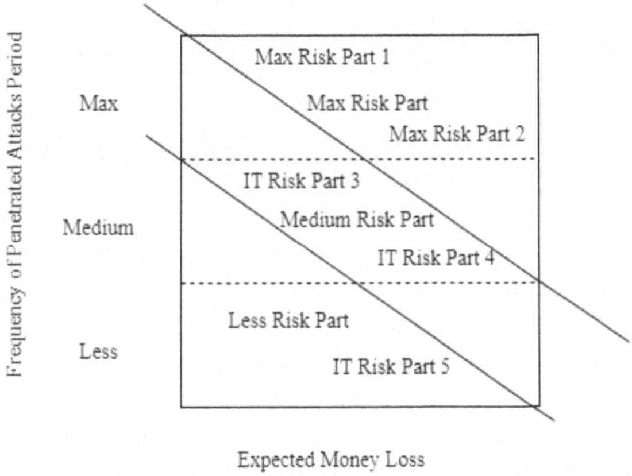

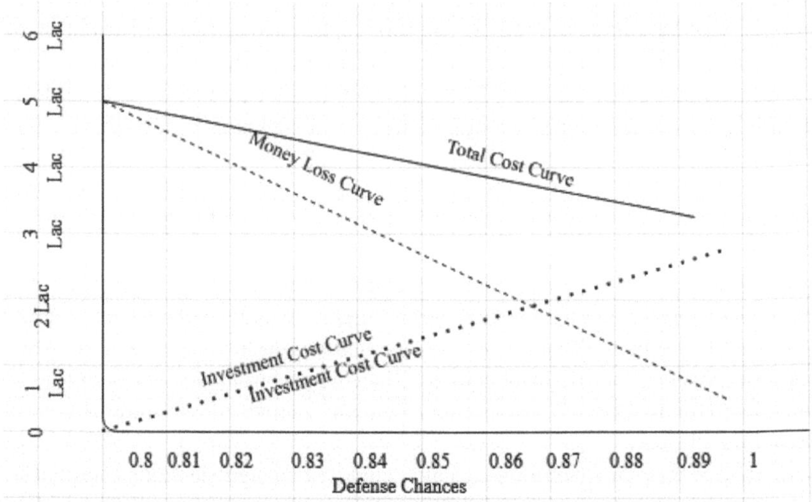

Fig. 7 Cost benefits analysis

Table 1 Design of IoT risks–threats

Design of IoT and asset–risks–threats		
layers	IoT devices and support	Risks and threat
Application layers	Audio-activated smartphone technological	IoT help–risks–threat–*v*: Audio-fixed devices' recordings might be distributed accidently to any individual or hotel staff, or they could be accessed by burglars and hotel attendants
Processing layer	A data center that is computerized, software-loaded, and virtual	IoT asset–risks–threat–*w*: Insiders and fraudsters utilize methods like hacking emails and breaking down passwords to access account information
Network layer	Personal LTE choices using a distributed antenna system (DAS)	IoT asset–risks–threat–*x*: Unprotected data in a message packet can be obtained by an attacker. The unprotected data's content could be manipulated by an intruder [31]
Perception layer	Wireless sensor networks (WSNs) with sensors and switches for automated lighting controls Climate control, airflow, and warming using networks of wireless sensors (NWS)	IoT help–risks–threat–*y*: Wi-Fi-enabled light bulbs are utilized in DDoS attacks. IoT help-risks-threats-: An invader take control to wireless device networks in order to perform harmful activities and send instructions to collect that has been deployed inside

4.3.1 Multiple IoT Asset–Vulnerabilities–Threats with Constraints

Table 1 shows the design of risk threats. One difficult question is how an organization can make the investment given the numerous financial and operational restrictions and the restricted organizational resources [30].

This research suggests a LPP methodology as money and mathematical restrictions to reduce the overall cyber cost of all IoT asset risks and threats working feature.

4.3.2 LPP-Based IoT Cyber Cost–Benefit Model

The IoT investing LPP method is denoted canonically.

Minimized cTa

Subjects to Ca $\leq b$ and $1 \geq a \geq 0$

here a talk about vector of main substitute for defense chances for IoT assets–risks–threats types, c and b are values of Constant, A is a value of coefficients, and $(\cdot)\ T$ is the opposite of matrix. The relation Ca $\leq b$ and $1 \geq a \geq 0$ are the limitations defining a conical value across which the desired task is to be minimized at the whole security cost. There are going to be five choices in accordance with the earlier illustration of the five asset-risks-attacks kinds, as well as a set of operational requirements and a set of non-negativity requirements [32].

4.4 Performance Layer

The decision to use a cyber solution and its designation by the Internet of things cyber evaluation tier start the Internet of things cyber performing actions. The discussion will be brief even though this layer is essential for managing risks because work generally follows that done at the risk assessment layer. Execution, control and evaluation, and continual enhancement are the three main activities at the IoT cyber performance layer [33].

4.4.1 Experimental Setup/Methodology

The creation, checking, deployments, new policy implement, teaching, and reader acceptability of IoT cyber technology are all included in the execution of the new IoT cyber architecture [34]. To evaluate and pick among the readily accessible security solutions included in the IoT cyber environment layer, organizations must define criteria for selection. The simplicity of use, usability, and utility of cyber control and monitoring systems must be considered while installing. Commercially available surveillance tools make it easier to build tracking and management systems, but they may also place restrictions on the developer's ability to customize the solution. The types of IoT applications and devices must also be considered when choosing the new IoT cyber infrastructure [35]. For example, some IoT devices are too small to put in calculation intensive security methods, so a gate-based design may be used to secure both the network as a whole and individual device.

4.4.2 Monitoring and Control

It is basic operations of preventing, identifying, and restoration are all carried out continuously to respond to various assaults. Data on the operation of the equipment, unusual user behavior, and unauthorized access [36] to data and applications are collected through detection as well as prevention activities. Real-time answers are provided via recovery actions. Risk Quantification parameters has been shown in Table 2.

4.4.3 Continuous Improvement

Important indicators of performance that the organization must monitor for continuous improvement include cyber-attack frequency and origin, result of infiltration [37] (such as revenue losses, paid fraud, fines, and data theft), and time taken to recover. The time of the work and the data that they are interested in is one distinction among control and surveillance and constant enhancement. Past information is used in perpetual improvement to find similarity and evaluate working overtime.

Table 2 Risk quantification parameter

Risk quantification parameter in Rs			
IoT asset–vulnerabilities–Threat	Financial loss function	Investment cost function	(Financial loss factor + Investment cost factor) Total cost factor
Type a	FL = Rs 480,000–Rs 4.000,000 * (a − 0.88)	IC = Rs 2,000,000 * (a − 0.88)	TC = Rs 480,000–Rs 2,000,000 * (a − 0.88)
Type b	FL = Rs 660,000–Rs 6.000,000 * (b − 0.89)	IC = Rs 2,500,000 * (b − 0.89)	TC = Rs 660,000–Rs 3,500,000 * (b − 0.89)
Type c	FL = Rs 500,000–Rs 5.000,000 * (c − 0.9)	IC = Rs 3,000,000 * (c − 0.9)	TC = Rs 500,000–Rs 2,000,000 * (c − 0.9)
Type d	FL = Rs 300,000–Rs 2,000,000 * (d − 0.85	IC = Rs 500,000 * (d − 0.85)	TC = Rs 300,000–Rs 1,500,000 * (d − 0.85)
Type e	FL = Rs 700,000–Rs 5.000,000 * (e − 0.86)	IC = Rs 2,000,000 * (e − 0.86)	TC = Rs 700,000–Rs 3,000,000 * (e − 0.86)

Surveillance and management, on the other hand, collect data in immediate fashion to monitor attacks and react to them instantly. Surveillance and management, on the other hand, collect data in immediate fashion to monitor attacks [38] and react to them instantly. Setting quantifiable performance targets and producing regular performance reports are essential components of ongoing enhancement tasks. To continuously develop, it is crucial to set objectives for success for each of IoT security initiatives using reference information gathered from the sector, industry, and rivals.

4.4.4 LPP Formula for IoT Cyber Resource Allocation

Consider that type a and c are not dependent and need a resource providing order separate, on the other hand type c, d, and e are related to network, and the resource providing of these three methods required to be minimum. Then, an LPP method of type c, d, and e proposed, and outcome is talked after executing of the method [39].

Execution for the LPP model:

(1) Decisions variable: Ballistic (Defense) probability for type c, d, and e, respect.
(2) The min ballistic probability: $c = 0.80$; $d = 0.80$; $e = 0.80$
(3) The max ballistic financial limitations to obtain the Ballistic probability of 0.84: chance (operational practicality constraint): all c, d, and $e = 0.89$

$$\text{Rs. } 2{,}000{,}000\, x + \text{Rs. } 400{,}000\, y \le \text{Rs. } 2{,}290{,}000,$$

$$\text{Rs. } 2{,}000{,}000\, x + \text{Rs. } 400{,}000\, y \ge \text{Rs. } 2{,}150{,}000,$$

$$\text{Rs. } 400{,}000\, y + \text{Rs. } 1{,}000{,}000\, z \le \text{Rs. } 1{,}350{,}000,$$

$$\text{Rs. } 2{,}000{,}000\, x + \text{Rs. } 1{,}000{,}000\, z \le \text{Rs. } 3{,}700{,}000,$$

$$\text{Rs. } 2{,}000{,}000\, x + \text{Rs. } 1{,}000{,}000\, z \ge \text{Rs. } 3{,}500{,}000,$$

Ballistic probability to a method feasible constraint between d and e: $d + e <= 0.81$.

Based on visualization, the LP model proposed is: Minimize GTC $= 6{,}155{,}000 - 1{,}000{,}000\, x - 7{,}500{,}000\, y - 2{,}000{,}000\, z$

Subject to

$$2{,}000{,}000\, x + 400{,}000\, y \le 2{,}290{,}000,$$

$$2{,}000{,}000\, x + 400{,}000\, y \ge 2{,}150{,}000,$$

$$400{,}000\, y + 1{,}000{,}000\, z \le 1{,}350{,}000,$$

$$2{,}000{,}000\, x + 1{,}000{,}000\, z \le 3{,}700{,}000,$$

$$2{,}000{,}000\, x + 1{,}000{,}000\, z \ge 3{,}500{,}000,$$

$$y + z \le 0.91$$

Table 3 Outcome of three cyber investing

Outcome of three cyber investing			
	Type 1	Type 2	Type 3 (LPP solution)
Ballistic probability	$a = 0.80$	$a = 0.84$	$a = 0.833333$
	$b = 0.80$	$b = 0.84$	$b = 0.88$
	$c = 0.80$	$c = 0.84$	$c = 0.83$
Money loss	$a =$ Rs. 400,000	$a =$ Rs. 100,000	$a =$ Rs. 223,335
	$b =$ Rs. 100,000	$b =$ Rs. 100,000	$b =$ Rs. 29,000
	$c =$ Rs. 400,000	$c =$ Rs. 100,000	$c =$ Rs. 240,000
	Total money loss: Rs 1,100,000	Total money loss: Rs 610,000	Total money loss: Rs 623,335
Investment cost	$a =$ Rs. 400,000	$a =$ Rs. 110,000	$a =$ Rs. 89,999
	$b =$ Rs. 24,000	$b =$ Rs. 44,000	$b =$ Rs. 55,000
	$c =$ Rs. 70,000	$c =$ Rs. 150,000	$c =$ Rs. 130,000
	Total investing cost: Rs. 104,000	Total investing cost: Rs. 315,000	Total investing cost: Rs. 294,999

Table 3 give Choice 3 is the least in case of cost to type c given to defense chances of 0.88, type e assigned to the defense chances of 0.833, and type c given to the defense chances of 0.83. The minimum total investing cost is Rs. 204,999. With the LPP method, Choice 3 gives a least total amount than Choice 2, with a lesser investing cost. This result shows the outcome of the LPP method in stuffed IoT cyber space providing method [40]. By altering the assumptions, multiple sensitivity tests can be carried out, giving management more confidence when choosing how to distribute funds using the LPP model. Each organization has a unique capacity for cyber defense, as suggested by the National Institute of Standards and Technology NIST's four Operational Tiers mode. The proposed LPP model can successfully consider the circumstances of each organization.

5 Conclusion

The IoT has been a basic system for autonomous vehicles, aerial vehicles, smart utilities, intelligent production, and intelligent healthcare. As more and more connected devices of all shapes and sizes are added to IoT networks, the potential security concerns rise significantly. When there is a lack of safety in IoT systems, burglars and attackers have a greater chance of gaining access to sensitive data and crucial infrastructure. Although, without some knowledge for managing IoT security hazards, it is extremely difficult for organizations to invest in and manage IoT cyber-security risks effectively. To mitigate cyber risks, this article examined IoT cyber-security technologies and knowledge. The four components of the IoT cyber risk control system that were provided in this study is the IoT digital environment, IoT digital structures, IoT digital risk evaluation, and IoT digital efficiency. This study intends to increase interest in the topic by filling a knowledge gap in the field of IoT risk management for cyber-security. For instance, present platforms do not provide managers with planning for resources tools. The result of this study shows the outcome of the LPP method in stuffed IoT cyber space. This result illustrates the outcome of the LPP method in crowded IoT cyber space providing method. By changing the assumptions, multiple sensitivity tests can be conducted, giving management more confidence when choosing how to distribute funds using the LPP model.

This shortcoming was addressed by the four-layer IoT cyber risk control framework that was recommended. Larger IoT systems with hundreds of decision factors, such as those used in intelligent production and smart transport, can be effectively scaled using the LP model.

Funding No institutions, organizations, or funding agencies have contributed any money to this research project.

Declaration of Competing Interest The authors of this study state that they have no known competing financial interests or personal ties that could have been perceived to have an impact on the work that was presented.

References

1. Kumar A, Parveen, Liu Y, Kumar R (eds) (2024) Handbook of intelligent and sustainable manufacturing: tools, principles, and strategies, 1st ed. CRC Press. https://doi.org/10.1201/9781003405870
2. Kumar A, Dogra N, Sarita, Bhatia S, Sidhu MS (eds) (2024) Handbook of intelligent and sustainable smart dentistry: nature and bio-inspired approaches, processes, materials, and manufacturing, 1st ed. CRC Press. https://doi.org/10.1201/9781003404934
3. Kumar A, Rajak DK, Kumar P, Rathee S (eds) (2024) Smart electric and hybrid vehicles: design, modeling, and assessment by industry 4.0 approaches, 1st ed. CRC Press. https://doi.org/10.1201/9781003502470
4. Kumar A, Rajak DK, Kumar P, Rathee S (eds) (2024) Smart electric and hybrid vehicles: fundamentals, strategies and applications, 1st ed. CRC Press. https://doi.org/10.1201/9781003495574
5. Dalal S, Kumar A, Lilhore UK, Dahiya N, Jaglan V, Rani U (2024) Optimizing cloud service provider selection with firefly-guided fuzzy decision support system for smart cities. Meas Sens 35:101294. https://doi.org/10.1016/j.measen.2024.101294
6. Kulshrestha R, Ramani P, Thakur P, Kumar A, Dogra N, Murthy KVSR, Nandan D (2024) An energy-efficient clustering approach for wireless sensor networks to reduce hot-spot effect and idle listening energy consumption. Math Model Eng Probl 11(7):1729–1742. https://doi.org/10.18280/mmep.110704
7. Naveena K, Krishnamoorthy M, Karuppiah N, Gouda PK, Hariharan S, Saravanan K, Kumar A (2024) Elevating sustainability with a multi-renewable hydrogen generation system empowered by machine learning and multi-objective optimization. Meas Sens 33:101192
8. Mohammadi M, Jamshidi S, Rezvanian A, Gheisari M, Kumar A (2024) Advanced fusion of MTM-LSTM and MLP models for time series forecasting: an application for forecasting the solar radiation. Meas Sens 33:101179

9. Sha K, Wei W, Yang TA, Wang Z, Shi W (2018) On security challenges and open issues in Internet of Things. Futur Gener Comput Syst 83:326–337
10. Ashraf I, Park Y, Hur S, Kim SW, Alroobaea R, Zikria YB, Nosheen S (2022) A survey on cyber security threats in iot-enabled maritime industry. IEEE Trans Intell Transp Syst 24(2):2677–2690
11. Rani S, Tripathi K, Arora Y, Kumar A (2022) Analysis of anomaly detection of malware using KNN. In: 2022 2nd International conference on innovative practices in technology and management (ICIPTM), vol 2. IEEE, pp 774–779
12. Malatji M (2023) Management of enterprise cyber security: a review of ISO/IEC 27001: 2022. In: 2023 International conference on cyber management and engineering (CyMaEn). IEEE, pp 117–122
13. de Azambuja AJG, Plesker C, Schützer K, Anderl R, Schleich B, Almeida VR (2023) Artificial intelligence-based cyber security in the context of industry 4.0—a survey. Electronics 12(8):1920
14. Rani S, Kumar A, Bagchi A, Yadav S, Kumar S (2021) RPL based routing protocols for load balancing in IoT network. J Phys Conf Ser 1950(1):012073
15. O'Neill M (2016) Insecurity by design: today's IoT device security problem. Engineering 2(1):48–49
16. Mukhopadhyay D (2016) PUFs as promising tools for security in internet of things. IEEE Design Test 33(3):103–115
17. Boeckl K, Boeckl K, Fagan M, Fisher W, Lefkovitz N, Megas KN, Scarfone K (2019) Considerations for managing Internet of Things (IoT) cybersecurity and privacy risks. US Department of Commerce, Gaithersburg, National Institute of Standards and Technology
18. Matheu SN, Hernandez-Ramos JL, Skarmeta AF (2019) Toward a cybersecurity certification framework for the Internet of Things. IEEE Secur Priv 17(3):66–76
19. Sohal AS, Sandhu R, Sood SK, Chang V (2018) A cybersecurity framework to identify malicious edge device in fog computing and cloud-of-things environments. Comput Secur 74:340–354
20. Yi S, Qin Z, Li Q (2015) Security and privacy issues of fog computing: a survey. In: Wireless algorithms, systems, and applications: 10th international conference, WASA 2015, Qufu, China, 10–12 Aug 2015. Proceedings 10. Springer International Publishing, pp 685–695
21. Medhane DV, Sangaiah AK, Hossain MS, Muhammad G, Wang J (2020) Blockchain-enabled distributed security framework for next-generation IoT: an edge cloud and software-defined network-integrated approach. IEEE Internet Things J 7(7):6143–6149
22. Tweneboah-Koduah S, Skouby KE, Tadayoni R (2017) Cyber security threats to IoT applications and service domains. Wirel Pers Commun 95:169–185
23. Rani S (2023) Big data query processing approach using MongoDB. In: 2023 3rd International conference on innovative practices in technology and management (ICIPTM). IEEE, pp 1–6
24. Gupta S, Rani S, Batra K (2020) Maximal security issues and threats protection in grid and cloud computing environment. Int J Adv Netw Appl 11(4):4367–4373
25. Nastase L (2017) Security in the internet of things: a survey on application layer protocols. In: 2017 21st international conference on control systems and computer science (CSCS). IEEE pp 659–666
26. Tekeoglu A, Tosun AŞ (2016) A testbed for security and privacy analysis of IoT devices. In: 2016 IEEE 13th international conference on mobile Ad Hoc and sensor systems (MASS). IEEE pp 343–348
27. Luo E, Bhuiyan MZA, Wang G, Rahman MA, Wu J, Atiquzzaman M (2018) Privacyprotector: privacy-protected patient data collection in IoT-based healthcare systems. IEEE Commun Mag 56(2):163–168
28. Khosravi-Farmad M, Ghaemi-Bafghi A (2020) Bayesian decision network-based security risk management framework. J Netw Syst Manage 28:1794–1819
29. Rea-Guaman AM, Mejía J, San Feliu T, Calvo-Manzano JA (2020) AVARCIBER: a framework for assessing cybersecurity risks. Clust Comput 23:1827–1843

30. Gordon LA, Loeb MP, Zhou L (2020) Integrating cost–benefit analysis into the NIST cybersecurity framework via the Gordon–Loeb Model. J Cybersecur 6(1)
31. Chakraborty A, Chandru V, Rao MR (2020) A linear programming primer: from Fourier to Karmarkar. Ann Oper Res 287:593–616
32. Georgescu TM, Iancu B, Zurini M (2019) Named-entity-recognition-based automated system for diagnosing cybersecurity situations in IoT networks. Sensors 19(15):3380
33. Alabady SA, Al-Turjman F, Din S (2020) A novel security model for cooperative virtual networks in the IoT era. Int J Parallel Program 48:280–295
34. Sun C (2012) Application of RFID technology for logistics on internet of things. AASRI Procedia 1:106–111
35. Babar S, Mahalle P, Stango A, Prasad N, Prasad R (2010) Proposed security model and threat taxonomy for the Internet of Things (IoT). In: Recent trends in network security and applications: 3rd international conference, CNSA 2010, Chennai, India, 23–25 July 2010. Proceedings 3. Springer Berlin Heidelberg, pp 420–429
36. Kumar H, Sarma D, Kar A (2008) Security threats in wireless sensor networks. IEEE Aerosp Electron Syst Mag 23(6):39–45
37. Riggs H, Tufail S, Parvez I, Tariq M, Khan MA, Amir A, Sarwat AI (2023) Impact, vulnerabilities, and mitigation strategies for cyber-secure critical infrastructure. Sensors 23(8):4060
38. Mazhar T, Irfan HM, Khan S, Haq I, Ullah I, Iqbal M, Hamam H (2023) Analysis of cyber security attacks and its solutions for the smart grid using machine learning and blockchain methods. Future Internet 15(2):83
39. Rajendra P, Kumari M, Rani S, Dogra N, Boadh R, Kumar A, Dahiya M (2022) Impact of artificial intelligence on civilization: future perspectives. Mater Today Proc 56:252–256
40. Rani S, Tripathi K, Arora Y, Kumar A (2022) A machine learning approach to analyze cloud computing attacks. In: 2022 5th International conference on contemporary computing and informatics (IC3I). IEEE, pp 22–26

Unlocking the Power of IoT, Machine Learning, and Cloud Computing: Case Studies in Smart Homes

Archna Goyal, Kamlesh Lakhwani, Bhavna Sharma, and Satyajeet Sharma

Abstract The integration of Cloud Computing, Internet of Things (IoT), and Machine Learning (ML) technologies is transforming smart homes. This chapter evaluates the impact of these technologies on energy efficiency, security, maintenance, and user experience. Through real-world case studies, we explore the implementation of personalized ambient control, predictive maintenance, intelligent security systems, and energy management systems. Our findings highlight significant benefits, including reduced energy consumption, enhanced security, proactive maintenance, and improved user comfort. We emphasize the critical role of scalable and secure cloud-based architectures and data-driven decision-making powered by ML algorithms. This study contributes scientifically by identifying key technological implementations and their impacts and outlining future research directions to enhance smart home environments.

Keywords Smart homes · Cloud computing · Internet of Things · Artificial intelligence · Machine learning

A. Goyal (✉)
K R Mangalam University, Gurugram, Haryana 122103, India
e-mail: goyalarchna511@gmail.com

K. Lakhwani · B. Sharma · S. Sharma
JECRC University, Jaipur, Rajasthan 303905, India
e-mail: kamlesh.lakhwani@jecrcu.edu.in

B. Sharma
e-mail: bhavna.sharma@jecrcu.edu.in

S. Sharma
e-mail: satyajeet.sharma@jecrcu.edu.in

A. Kumar et al. (eds.), *Trends in Mechatronics Systems*, Emerging Trends in Mechatronics, https://doi.org/10.1007/978-981-97-9108-8_8

1 Introduction

Cloud computing, a cutting-edge technology, enables the provision of various computer services, including networking, servers, analytics, storage, software, databases, and intelligence over the internet [1]. Its characteristics are shown in Fig. 1. This integration has transformed how households manage their electronic settings, enabling real-time adaptation and control of resources based on individual needs from any location with an internet connection [2]. Cloud computing serves as the primary hub for managing and processing data from numerous connected devices [3], allowing homeowners to control temperature, lighting, security systems, and appliances remotely, thus enhancing convenience and energy efficiency [4].

Machine learning, a subfield of artificial intelligence, allows computers to process data without explicit programming [5]. It enhances smart homes by providing predictive analytics, natural language processing, image recognition, recommendation systems, anomaly detection, and healthcare applications [6]. These capabilities contribute to improved comfort, efficiency, and safety in smart homes through tailored automation, adaptive energy management, and proactive security measures [7].

The Internet of Things (IoT) is a network of connected physical objects, such as vehicles, appliances, and other items embedded with sensors, software, and network connectivity [8]. This technology facilitates the direct integration of physical systems into computer systems, enhancing efficiency, providing economic benefits, and reducing human intervention [9]. IoT is widely used in smart homes, healthcare, industrial automation, and agriculture, creating smart environments and improving

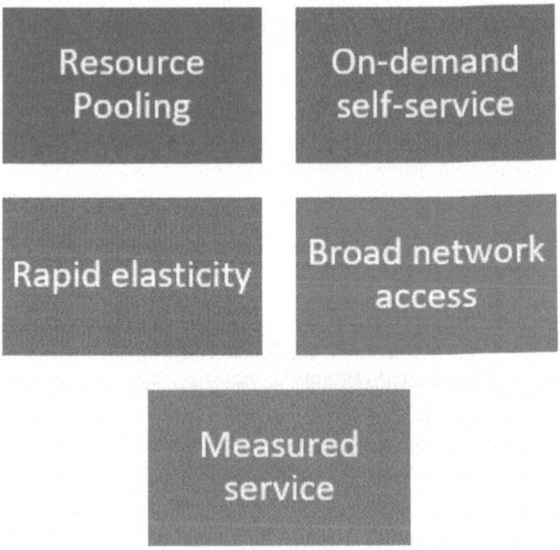

Fig. 1 Characteristics of cloud computing

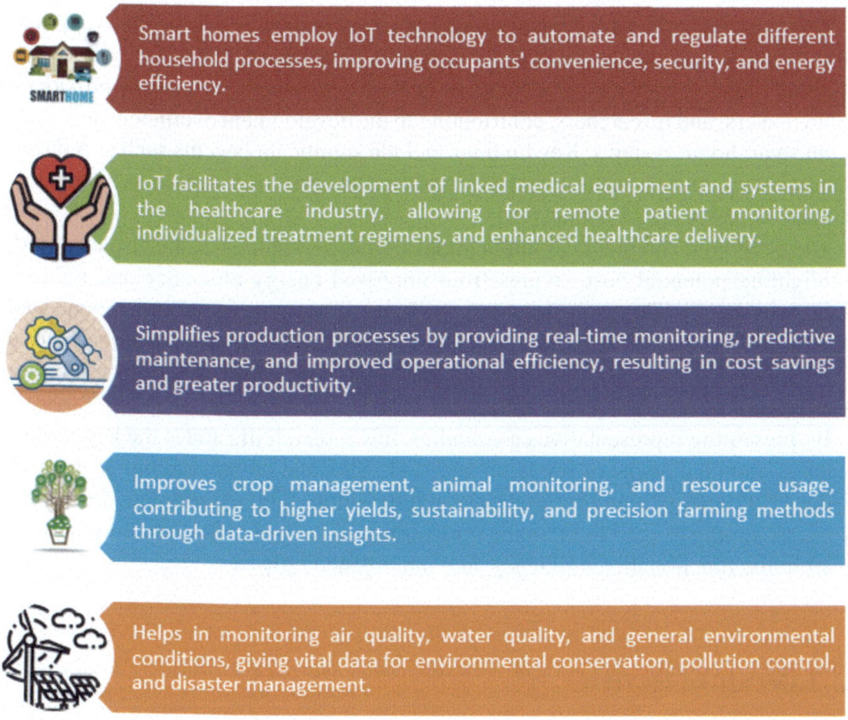

Fig. 2 Domains of IoT

decision-making processes based on data-driven insights [10]. Some of its applications are shown in Fig. 2, which illustrates the various domains where IoT technology is applied, including smart homes, healthcare, industrial automation, and agriculture.

Smart homes leverage the integration of cloud computing, IoT, and ML to create environments that enhance occupant comfort, safety, and energy efficiency [11]. These homes consist of interconnected devices and systems that enable automation, remote control, and intelligent decision-making [12]. The robust infrastructure of cloud computing ensures necessary storage, processing, and computing capabilities, while ML algorithms analyze data from smart home devices to deliver predictive and intelligent functionalities [13, 14].

This study addresses the following key research questions.

- How do IoT, ML, and cloud computing technologies contribute to smart home environments?
- What are the specific benefits and challenges associated with their implementation?
- How does cloud computing enhance the functionality of smart home systems?
- What are the critical benefits of using machine learning in smart home environments?

By examining real-world case studies, this research evaluates the practical applications, benefits, and challenges of cloud computing, IoT, and ML in smart homes [15]. The study aims to provide valuable insights for homeowners, technology providers, policymakers, and researchers, contributing to the development of efficient and intelligent smart home systems. Key findings include significant benefits such as reduced energy consumption, customized ambient settings, proactive device maintenance, and advanced security systems [16].

The economic and practical implications of these technologies are also explored, highlighting potential cost savings from improved energy efficiency and reduced maintenance [17]. The study discusses the balance between initial investments and long-term benefits, emphasizing the importance of scalable and secure cloud-based architectures in managing large data volumes and enabling real-time processing for informed decision-making and remote access [18].

By presenting representative case studies, this research illustrates the key results, benefits, and challenges of implementing IoT, ML, and cloud computing in smart homes [19]. It also discusses future research directions and opportunities for further progress, aiming to shed light on the transformative impact of these technologies on smart home functionality, efficiency, and intelligence [20].

2 Literature Review

Botta et al. [1] provide insights on cloud computing and IoT integration, stressing the benefits and pitfalls of this fusion. Singh et al. [2] explore into the realm of edge computing in IoT, providing case studies and an assessment of edge computing's practical applications. Similarly, Mohammadi et al. [5] concentrate on the applying machine learning techniques in IoT, outlining and assessing the important methodologies employed in this junction. Kulshrestha et al. [8] provides survey of IoT technology and applications, providing an in-depth review of the topic.

Rani et al. [3] undertake a thorough study of the security challenges related with cloud computing in the context of smart grid systems, putting light on significant concerns in the sector. Kaur and Kaur [9] provide a detailed analysis of how cloud computing and machine learning are integrated, stressing the potential benefits and problems of this combination. Xu et al. [21] investigate data privacy issues in smart homes, offering an assessment of the current landscape via case studies and surveys.

Phogat et al. [4] focus on analyzing the key technologies involved in smart home system architecture, offering valuable insights into the foundational elements of such systems. Majeed et al. [22] present a comprehensive review of energy management in smart homes, highlighting the various strategies and technologies employed for efficient energy utilization.

Naveena et al. [6] give a thorough analysis of machine learning's function in IoT data processing, emphasizing the many use cases and issues in this field. Malche and Maheshwary [10] investigate the practical uses of IoT in the construction of smart home systems, offering insights into the technology's prospective deployment. The integration of cloud computing, IoT, and machine learning in smart environments has also been explored in various domains, such as agriculture, where IoT-based systems have demonstrated significant potential for efficient resource management and automation [14]. Lakhwani et al. [23] examine adaptive workflow scheduling in cloud environments, offering insights into optimization approaches that could benefit the deployment of scalable and efficient smart home systems. Ghayvat et al. [11] explore the chances and problems of adopting IoT in smart homes and buildings, stressing the potential advantages and constraints of this integration. Cook et al. [13] introduce the CASAS smart home system, providing an in-depth examination of its design and operation. Soliman et al. [24] investigate how IoT may be integrated with cloud computing and web services in smart home systems, offering a thorough examination of the benefits and drawbacks. Furthermore, Machorro-Cano et al. [25] exhibit an energy-efficient smart house system with the help of machine learning and big data, demonstrating the real-world implications of implementing advanced energy management technologies in smart homes.

Dalal et al. [7] give a detailed evaluation of the applications of deep learning algorithms for face and activity detection, emphasizing the significance of deep learning in strengthening security in smart homes. Xiao [16] investigates the use of machine learning techniques for energy monitoring in smart home systems, emphasizing the need of data-driven approaches for optimal energy management. Djumanazarov et al. [26] provide an overview and study of smart home system architecture based on the Internet of Things. Furthermore, Sharma et al. [27] present an optimization-driven machine learning approach for improving of security smart home through the use of 5G technology, providing insights into the potential benefits of adopting sophisticated network technologies for security enhancement. Yang [28] explores the usage of cloud computing technology in the creation of smart home models, offering useful insights into the practical consequences and advantages of cloud computing in this context.

Saleem et al. [12] present a detailed analysis of machine learning applications in smart home systems, stressing the potential benefits and problems of this integration. Priyadarshini et al. [15] demonstrate the practical implications of data-driven models for energy management by introducing an ensemble approach for machine learning that predicts energy use in smart homes. Sinha et al. [17] offer a smart house automation system based on IoT and cloud computing, highlighting the potential beneficial effects and limitations of combining these technologies for smart home automation. Furthermore, Peng et al. [29] investigate a deep learning-based smart home system, demonstrating the potential benefits and limitations connected with deep learning-based smart home systems. Finally, Shafi et al. [30] analyze how a machine-learning approach may be used to IoT-based smart homes, providing insights into its practical uses and consequences. The summary of the literature review is given in Table 1, which provides a comparative study of the literature reviewed.

These studies add to the growing body of knowledge in the field by emphasizing the critical intersections between IoT, machine learning, cloud computing, and smart home technologies, as well as emphasizing the importance of addressing

Table 1 Comparative study of literature review

S. No.	Paper title	Authors	Year	Focus of study	Findings	Research gap
1	Integration of Cloud Computing and Internet of Things: A Survey	Botta, A., De Donato, W., Persico, V., and Pescapé, A	2016	Cloud computing and IoT integration	Analysis of the integration of cloud computing and IoT	Potential economic and regulatory implications of integrating cloud computing with IoT are not specified
2	A Survey of Edge Computing-Based Internet of Things	Yao, L., Wu, Z., Tang, J., and Guo, Y	2017	Edge computing in IoT	Assessment of edge computing applications in IoT	Scalability and interoperability challenges are not explored
3	Security and Privacy in Cloud Computing for Smart Grid Systems: A Comprehensive Review	Yang, Y., Li, Y., and Luo, J.	2017	Cloud security for smart grid systems	Comprehensive review of security issues in cloud computing for smart grid systems	Assessment of the legal and ethical implications of cloud security not done
4	Research on Smart Home System Architecture and Key Technologies	Wang, Y., Zhang, Y., and Zhang, J	2016	Smart home architecture	Analysis of key technologies in smart home systems	Lack of comprehensive evaluation of usability and user experience
5	A Survey on Machine Learning Techniques in Internet of Things	Le, T., and Wang, C	2019	ML techniques in IoT	Review of ML techniques applied in IoT	Lack of practical implementation challenges and real-world implications of various ML techniques in IoT
6	Machine Learning for Internet of Things Data Analysis: A Survey	Mahdavinejad, M. S., Rezvan, M., Barekatain, M., Adibi, P., Barnaghi, P., and Sheth, A. P	2018	Machine learning in IoT data analysis	Comprehensive overview of machine learning applications in IoT data analysis	Ethical implications and potential biases connected with machine learning applications in IoT data processing were not properly explored

(continued)

Table 1 (continued)

S. No.	Paper title	Authors	Year	Focus of study	Findings	Research gap
7	Smart Home: Deep Learning as a Method for Machine Learning in Recognition of Face, Silhouette and Human Activity in the Service of a Safe Home	Vardakis, G., Tsamis, G., Koutsaki, E., Haridimos, K., and Papadakis, N	2022	Deep learning in smart home security	Exploration of deep learning applications for enhancing security in smart homes	Data protection issues associated with the use of deep learning for face and activity recognition
8	The Internet of Things: A Survey	Atzori, L., Iera, A., and Morabito, G	2010	Overview of IoT technologies	Overview of IoT technologies and applications	Insufficient exploration of the potential challenges and limitations in the widespread implementation of IoT technologies
9	Cloud Computing and Machine Learning: A Comprehensive Review	Kaur, R., and Kaur, A	2020	Integration of cloud computing and ML	Comprehensive review of the integration of cloud computing and ML	Assessment of the legal and ethical implications of cloud security are insufficiently explored
10	Internet of Things (IoT) for Building Smart Home Systems	Malche, T., and Maheshwary, P	2017	Smart home system	Exploration of IoT applications in building smart home systems	Discussion of the actual challenges and limitations to deploying IoT in smart home systems is limited
11	Internet of Things for Smart Homes and Buildings: Opportunities and Challenges	Ghayvat, H., Liu, J., Babu, A., Alahi, E. E., Gui, X., and Mukhopadhyay, S. C	2015	IoT applications in smart homes and buildings	Identification of opportunities and challenges for implementing IoT in smart homes and buildings	Cybersecurity threats and vulnerabilities related to the expanding use of IoT in smart homes and buildings are not discussed properly

(continued)

Table 1 (continued)

S. No.	Paper title	Authors	Year	Focus of study	Findings	Research gap
12	Smart Homes Powered by Machine Learning: A Review	Saleem, A. A., Hassan, M. M., and Ali, I. A	2022	Machine learning applications in smart home systems	Review of the applications and benefits of machine learning in smart home systems	Inadequate evaluation of the social and cultural consequences of incorporating machine learning into smart home systems, such as possible biases and equality problems
13	CASAS: A Smart Home in a Box	Cook, D. J., Crandall, A. S., Thomas, B. L., and Krishnan, N. C	2013	Smart home system architecture	Overview of the CASAS smart home system and its architecture	Inadequate assessment of the CASAS smart home system's long-term usefulness and adaptation in diverse environments
14	A Machine-Learning Ensemble Model for Predicting Energy Consumption in Smart Homes	Priyadarshini, I., Sahu, S., Kumar, R., and Taniar, D	2022	Machine learning for energy consumption prediction in smart homes	Development and evaluation of a machine-learning ensemble model for predicting energy consumption in smart homes	Upgrade problems related to the use of machine learning ensemble models for energy prediction in smart homes
15	Machine Learning in Smart Home Energy Monitoring System	Xiao, G	2021	Machine learning for energy monitoring in smart homes	Examination of machine learning techniques for energy monitoring in smart home systems	User experience and usability issues associated with the use of machine learning algorithms for energy monitoring in smart home systems
16	Smart Home Automation Using IoT with Cloud Computing	Sinha, S., Raj, T., and Kumar, S	2023	IoT-based smart home automation with cloud computing	Proposal and demonstration of smart home automation utilizing IoT and cloud computing	Possible regulatory and policy ramifications of the broad deployment of IoT and cloud computing in smart home automation are not discussed

(continued)

Table 1 (continued)

S. No.	Paper title	Authors	Year	Focus of study	Findings	Research gap
17	A Survey on Data Privacy in Smart Homes	Xu, Y., Khorasgani, H., and Ma, J	2019	Data privacy in smart homes	Evaluation of data privacy issues in smart homes	Social and psychological implications of data privacy concerns are not properly discussed
18	Application of Cloud Computing Technology in Smart Home Modeling Design	Yang, S	2022	Cloud computing in smart home design	Application of cloud computing technology for designing smart home models	Possible data governance and ownership concerns that may arise as a result of the usage of cloud computing technologies in the creation of smart home models
19	An Overview of IoT-Based Architecture Model for Smart Home Systems	Djumanazarov, O., Väänänen, A., Haataja, K., and Toivanen, P	2022	Architecture model for IoT-based smart home systems	Description and analysis of an IoT-based architecture model for smart home systems	Economic and financial consequences of using IoT-based architectural models for smart home systems are not explored
20	IoT Based Smart Home: A Machine Learning Approach	Shafi, M. K. I., Sultan, M. R., Rahman, S. M. M., and Hoque, M. M	2021	Machine learning approach for IoT-based smart homes	Utilization of a machine learning approach for IoT-based smart home systems	Standards concerns associated with the use of machine learning algorithms in IoT-based smart home systems
21	Smart Home Automation Using IoT with Cloud Computing	Sinha, S., Raj, T., and Kumar, S	2023	IoT-based smart home automation with cloud computing	Proposal and demonstration of smart home automation utilizing IoT and cloud computing	Possible regulatory and policy ramifications of the broad deployment of IoT and cloud computing in smart home automation are not discussed

security, privacy, and energy management concerns for the continued advancement and adoption of smart home systems.

3 Methodology

Through the comparison of case studies, this study investigates the different applications of cloud computing, machine learning, and the Internet of Things in smart homes the technique entails analyzing the use of personalized ambient control, predictive maintenance, intelligent security systems, and energy management systems in smart home environments, as outlined in the flow chart for methodology shown in Fig. 3. The main technologies, uses cases, and advantages of integrating various technologies are assessed in this study. It also explores the difficulties and constraints noted in the case studies and points out areas that might use additional investigation and advancement.

The case studies were selected through a comprehensive review of existing literature, industry reports, and expert consultations. The criteria for selection included:

1. **Relevance to Smart Home Technologies**: The case studies needed to demonstrate clear applications of cloud computing, ML, and IoT within a smart home context.
2. **Diversity in Technological Applications**: To ensure a broad understanding, the case studies encompassed various applications such as energy management, intelligent security systems, predictive maintenance, and personalized ambient control.

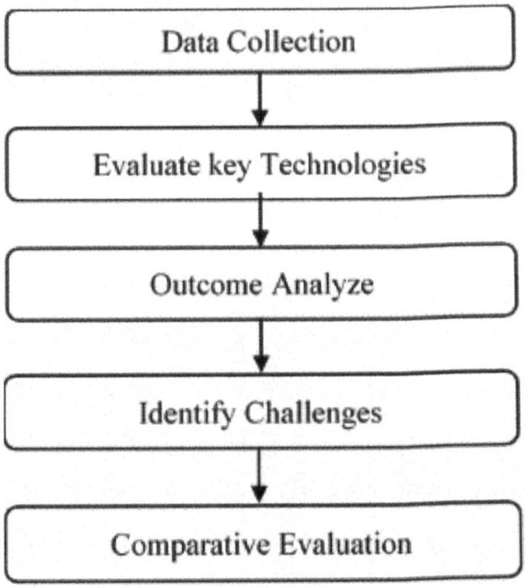

Fig. 3 Flowchart for methodology

3. **Availability of Detailed Data**: Only case studies with sufficient data on implementation, outcomes, and challenges were included for thorough analysis.

Each case study was analyzed using a systematic approach that involved:

- **Data Collection**: Gathering detailed information on the technological setup, data flow, and integration methods from IoT devices to cloud infrastructure.
- **Technology Assessment**: Evaluating the specific technologies used, such as types of IoT devices, ML algorithms, and cloud services.
- **Outcome Analysis**: Assessing the results of each case study, including improvements in efficiency, security, user experience, and energy consumption.
- **Challenge Identification**: Identifying any difficulties and limitations encountered during implementation and operation.
- **Comparative Evaluation**: Comparing the outcomes across different case studies to highlight the benefits and drawbacks of each technological integration.

4 Smart Homes and the Internet of Things (IoT)

The term "Internet of Things" (IoT) describes networks of electronically interconnected devices, sensors, and physical things that facilitate data collection and exchange [1]. IoT technology enables different devices such as lighting systems, security cameras, thermostats, and appliances to connect with one another and with homeowners in smart homes context [8]. This connected network forms the basis of the smart home, enabling automation, remote control, and intelligent decision-making [10].

4.1 IoT Architecture and Components

IoT architecture for smart homes typically includes three main components: a device or sensor, a connectivity layer, and a data processing layer, as shown in Fig. 4. Devices and sensors include an extensive range of physical gadgets, such as Motion sensors and smart meters, as well as smart speakers and wearables [9]. Connectivity layer facilitates communication and data transfer between these devices using various wireless protocols such as Wi-Fi, Bluetooth, and Zigbee [10]. The data processing layer consists of cloud-based platforms or edge devices that receive, store, and analyze collected data to derive insights and trigger actions [31].

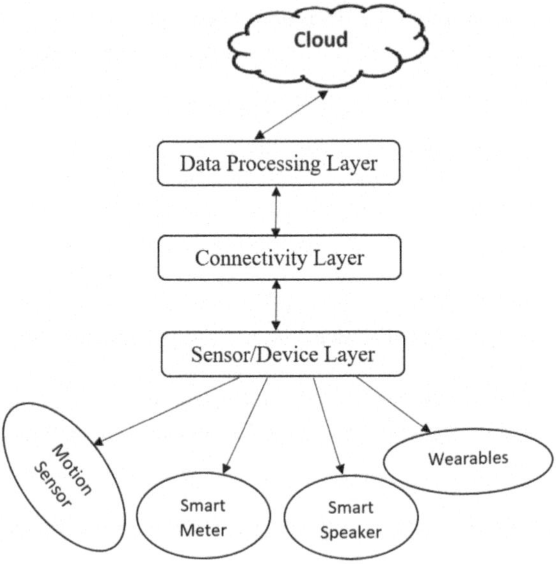

Fig. 4 IoT architecture for smart homes

4.2 IoT Application in Smart Home

IoT technologies provide a variety of uses and applications in the smart home context. IoT enables homeowners to automate a variety of tasks and processes, including adjust temperature settings, control lighting systems, and manage devices remotely [1]. Various IoT applications are shown in Fig. 5. This automation will improve the comfort, energy efficiency and convenience of smart homes. IoT devices and sensors can monitor and optimize smart home energy consumption. By collecting data on energy usage patterns, homeowners can identify energy savings opportunities and make informed energy usage decisions [8]. IoT-based security systems offer advanced surveillance and surveillance capabilities such as smart locks, motion sensors, and video cameras. These systems enhance home security, alert homeowners to potential threats, and enable remote monitoring and control [10]. IoT technology can support the elderly and disabled by providing assistive devices and remote monitoring systems. These systems monitor health parameters, identify emergencies, and enable caregivers to provide remote assistance [11].

4.3 Challenges and Security Considerations

While IoT technology brings great benefits to the smart home, it also poses some challenges and security issues. IoT devices create enormous amounts of data, raising privacy and data security concerns. Unauthorized access to personal data and potential data breaches are major concerns that must be addressed through strong security

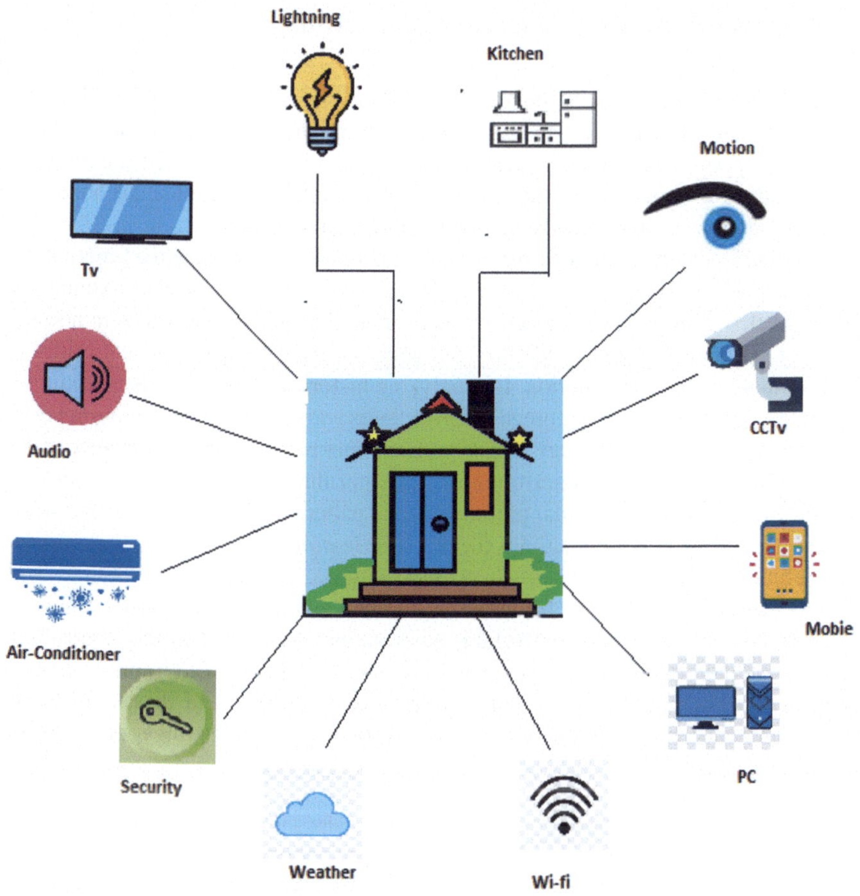

Fig. 5 IoT applications in smart homes

measures and data encryption [13]. As the number of IoT devices and platforms becomes available, interoperability and standardization will become key issues. Industry-wide standards and protocols are needed to ensure seamless communication and integration between devices from different manufacturers [16]. IoT devices rely heavily on a stable internet connection. Connectivity issues, network disruptions, or device failures can compromise the reliable functioning of smart home systems, requiring redundancy measures and backup solutions [17]. The complexity of IoT systems and interfaces can pose challenges for homeowners in terms of setup, configuration, and user experience. Designing a user-friendly interface and providing clear instructions is essential to ensure a positive user experience [31].

5 Machine Learning (ML) for Smart Homes

The goal of the artificial intelligence (AI) discipline of machine learning is to develop algorithms and models that can learn from data, predict needs, and react without explicit programming. Machine learning algorithms analyze the vast amounts of data produced by Internet of Things (IoT) devices in the context of smart homes to uncover patterns, identify trends, and make informed decisions.

Machine learning technology plays a prominent role in increasing the functioning and intelligence of smart homes. Machine-learning algorithms can analyze patterns in data collected from various sensors to detect anomalies and deviations from normal behavior. This allows you to identify potential security breaches, device failures, or abnormal energy consumption. By analyzing historical data and sensor readings, machine learning models can forecast device faults and maintenance requirements in advance. This proactive strategy assists homeowners in planning maintenance tasks and avoiding unanticipated interruptions [2]. Algorithms that use machine learning are able to adjust to the particular preferences and patterns of individual homeowners, as depicted in Fig. 6, which outlines the machine learning approach in smart homes. By continuously analyzing data such as usage patterns, temperature preferences and lighting preferences, systems for smart homes can automatically adjust settings to generate a personalized and comfortable environment. Machine learning algorithms can analyze patterns in energy usage and environmental factors to improve energy usage in smart homes. Analyzing both historical and real-time data, these algorithms can recommend energy efficiency actions, adjust thermostat settings, or optimize device usage [6].

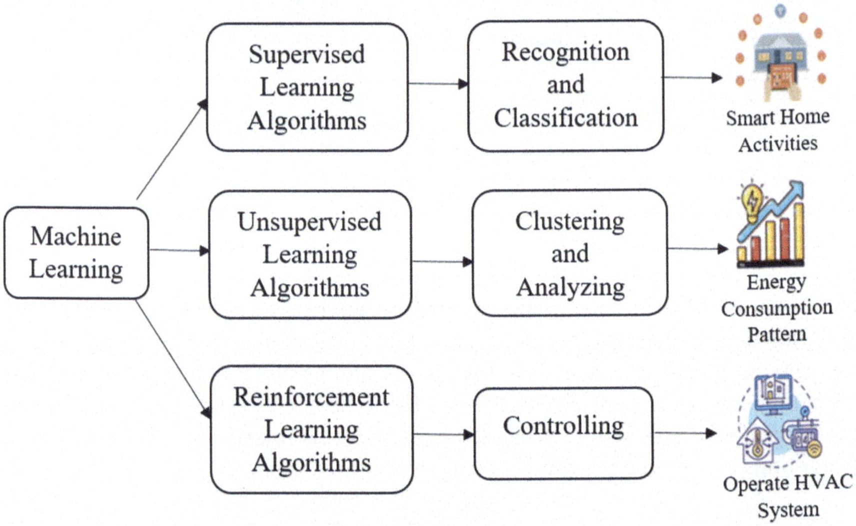

Fig. 6 Machine learning approach in smart homes

Smart homes can utilize supervised learning algorithms to recognize and categorize activities [6]. By using labeled data such as sensor readings and video footage to train the algorithm, it can accurately identify activities such as cooking, cleaning, or watching TV. This enables personalized automation and intelligent assistance [12]. In smart homes, supervised learning is useful to identify unusual behaviors and occurrences. By training algorithms with tagged data representing normal activity, you can detect anomalies that can indicate security breaches, device malfunctions, or unusual energy consumption [19].

Unsupervised learning algorithms is useful to analyze smart home energy utilization patterns [6]. By grouping similar energy consumption profiles, the algorithm identifies energy saving opportunities, identifies anomalies, and provides insights for optimizing energy efficiency [12]. Unsupervised learning can be used to derive smart home occupancy patterns. By analyzing sensor data such as door/window sensors and motion sensors, algorithms can identify occupancy patterns, identify outliers, and help optimize heating, cooling, and lighting based on occupancy [20].

In smart homes, reinforcement learning algorithms can learn how to operate HVAC (heating, ventilation, and air conditioning) systems [11]. The algorithm interacts with the environment (temperature sensors, weather forecasts, etc.) and learns how to make the best decisions to minimize energy consumption while maintaining a desired level of comfort [12]. Users can receive personalized suggestions based on their tastes and past data using reinforcement learning. The algorithm can learn how to suggest energy-efficient behaviors, lighting settings, and device usage patterns that are tailored to each user's needs and preferences [19].

In smart homes, ML (machine learning) models can be applied to various aspects such as predictive maintenance, energy optimization, and security. ML models can analyze sensor data from devices and appliances within a smart home to identify patterns indicative of potential faults or malfunctions [11]. By training the models on historical data, they can predict when a device or system is likely to fail, allowing for proactive maintenance or repair. They can continuously monitor the health of devices and systems, tracking parameters such as temperature, vibration, or power consumption. By learning normal operating conditions, the models can detect deviations and provide early warnings or alerts for maintenance needs [12]. ML models can forecast electricity demand within a smart home based on historical usage patterns, weather conditions, and other relevant factors. This enables optimized scheduling of energy-consuming devices and helps homeowners make informed decisions regarding energy usage [15]. They can analyze historical energy consumption data to identify usage patterns and provide insights on energy-saving opportunities. By understanding peak usage periods, inefficient appliances, or wasteful habits, homeowners can adjust their behaviors and optimize energy consumption [21]. ML models can learn normal behavioral patterns in a smart home by analyzing sensor data, user interactions, and device activities. This allows you to detect anomalies and suspicious behavior that can indicate security threats, such as unauthorized access attempts and unusual activity [32]. Tagged data can be trained to detect different types of security threats and intrusion patterns. Models are able to identify and categorize possible security breaches and instances of illegal access by examining data from security cameras, motion sensors, and other devices [22].

6 Cloud Computing for Smart Homes

Cloud computing is the distribution of computer resources to be used as a service on the Internet, such as processing power, storage and software programs for use under pay per use [3]. Cloud computing services provide a centralized platform for storage, management, and analysis of vast amounts of data generated by Internet of Things devices in the context of smart homes [5].

6.1 Cloud-Based Architectures and Applications for Smart Homes

Cloud computing offers various services that benefit smart homes. Cloud storage enables homeowners to securely store and access their data from anywhere at any time. This ensures that data collected by IoT devices, such as sensor readings or security footage, can be easily retrieved and analyzed [3]. Cloud services make it easier for homeowners to connect and manage their home devices and systems remotely. They allow homeowners to monitor and manage security systems, adjust settings, and receive real-time notifications on the go [5]. Cloud computing offers scalability, allowing smart home systems to handle increasing amounts of data and device connectivity. Homeowners can easily add or remove devices without requiring major infrastructure upgrades [6]. Cloud platforms provide powerful computational capabilities for processing and analyzing IoT data. Leverage machine learning algorithms, analytical tools, and real-time processing engines to identify patterns, derive valuable insights, and make data-driven decisions [16].

Cloud computing allows smart homes to offload data storage and processing tasks to remote servers, offloading the computational load on local devices [18] as shown in Fig. 7. In order to ensure data redundancy and reduce the risk of loss, information collected by IoT devices will be securely transferred and stored in the cloud. Cloud-based data processing enables advanced analytics, artificial intelligence, machine learning capabilities [19]. Homeowners can leverage cloud platforms to train machine learning models, run complex algorithms, and perform in-depth data analysis for more accurate predictions and intelligent decision making [33].

Cloud computing brings scalability and flexibility to the smart home to accommodate an increasing number of devices and growing data complexity [6]. The cloud-based platform handles data influx from multiple sources, supports real-time data processing, and can scale resources as needed to meet changing demands [16].

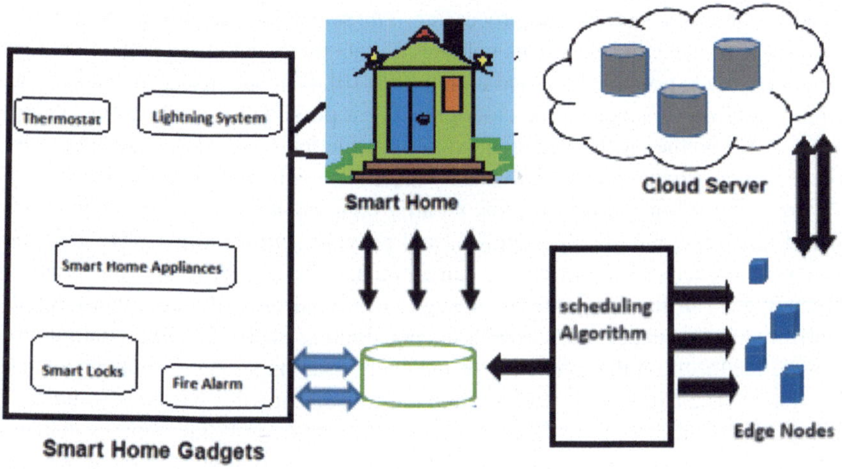

Fig. 7 Cloud-based architectures for smart homes

Because of the freedom that cloud computing gives, homeowners may select from a large range of services and apps that are suited to their unique needs [18]. You may construct a customized smart home ecosystem by selecting and integrating several cloud-based solutions for security, energy management, home automation, and health monitoring [27].

Cloud computing offers great benefits, but privacy and security issues must be addressed. Storing data in the cloud raises questions of privacy, ownership, and control. Homeowners must ensure data is encrypted, access is restricted, and privacy regulations are adhered to [19]. Cloud providers must provide robust security measures to prevent unauthorized access, service interruptions and data breaches. This includes measures such as encryption, access control, security of the network, and regular audits to ensure the integrity and confidentiality of smart home data [33].

7 Case Studies

7.1 Energy Management in Smart Home with Machine Learning, IoT, and Cloud Computing

Energy management and optimization in smart homes is a compelling case study. Smart plugs, smart lighting systems, smart appliances, Energy monitoring systems and smart thermostats are internet-connected devices that collect data in real-time on occupancy, weather, and other pertinent factors as well as trends of energy use. This data are sent to a central hub or gateway that acts as a bridge between your IoT devices and your cloud infrastructure [4]. Energy data are gathered and transferred to the cloud for analysis, storing, and processing. The scalability and processing capacity required to manage the enormous volumes of data produced by the numerous smart homes are made possible by cloud computing [10]. Data are managed and kept in a safe, centralized manner on the cloud, making it possible to save historical energy usage data together with other contextual information for analysis and future use. These gathered data sets are analyzed by machine learning algorithms [19].

These algorithms derive insights, identify patterns, predict energy usage trends, identify energy savings opportunities, and provide actionable insights for homeowners. For example, smart homes can automatically modify temperature settings depending on utilization patterns or suggest optimal energy usage plans, significantly reducing energy waste and lowering running costs [22]. We continuously provide feedback on the efficiency of the energy-saving strategies implemented in the system. The ML model is refined and improved, using this feedback resulting in more accurate predictions and better energy optimization over time. Energy management for smart home using cloud computing, machine learning, and the Internet of Things is an innovative approach to optimize energy consumption and improve efficiency in residential buildings. This integrated system uses the capabilities of IoT devices, ML algorithms, and cloud-based infrastructure to monitor, analyze, and control energy consumption in smart home environments [27].

7.2 Intelligent Security Systems for Smart Homes Leveraging ML Algorithms

Another important case study is the application of cloud computing, machine learning, and IoT (Internet of Things) to intelligent security systems for smart homes. Smart home security systems use various IoT devices such as security cameras, motion detectors, door/window sensors, and smart locks. These Internet-connected gadgets gather data in real time regarding environmental factors and security incidents [5]. IoT devices continuously collect data about events such as motion detection, door/window opening and closing, and camera feeds. This information is transmitted to a central hub or gateway, which serves as a link for communication between your IoT devices and your cloud infrastructure. Security data is collected and moved to the cloud for analysis, processing, and storing. The computer capacity and scalability required to process the data generated by several smart homes are provided by cloud computing [7]. This involves archiving video footage, sensor data, and security event logs for analysis, review, and reference in the future [11]. ML algorithms are used to find patterns, abnormalities, and possible dangers in security data. Machine learning algorithms can learn from previous data to identify regular activity patterns and variations that may signal security breaches or dangers. Based on specified rules and learnt patterns, ML models may continually analyze security data and provide real-time warnings and notifications. This allows homeowners and security staff to be alerted in real time of security problems, unauthorized entry attempts, or suspicious activities [12]. Smart home security systems may be accessed and controlled remotely thanks to cloud-based technology. Homeowners may view live video feeds, get security warnings on their mobile devices, and manage security equipment like locks and alarm systems remotely. Smart security systems can be integrated with other smart home devices and services. For example, it can connect to smart lighting systems to automatically turn on lights in the event of a security breach, or integrate with voice assistants to execute voice-activated security commands. Combining IoT, machine learning, and cloud computing, smart home security systems provide enhanced security capabilities, real-time monitoring, and intelligent threat detection capabilities. These systems will allow homeowners to better manage the security and peace of mind in their homes [13].

7.3 Predictive Maintenance of Smart Home Devices Using Cloud-Based Analytics

This case study focuses on the application of cloud-based analytics for predictive maintenance of smart home devices. By collecting and analyzing data from his IoT devices in smart homes, cloud-based analytics can predict potential errors and malfunctions before they occur [5]. This enables proactive maintenance, minimizes downtime, and optimizes device performance. Collected data is stored securely in the cloud, providing scalability and computing power for complex analysis [6]. Machine learning models are developed and trained based on historical data to identify patterns and anomalies related to device performance and potential failures. When anomalies are detected, alerts are generated to notify homeowners and maintenance personnel [19]. Predictive maintenance analytics also provides insights into efficient maintenance planning and resource allocation. Continuous learning and updating of machine learning models ensure accuracy and effectiveness over time. Ultimately, using cloud-based analytics for predictive maintenance will improve device reliability, reduce costs, and improve the overall user experience in smart homes [33].

7.4 Personalized Ambient Control in Smart Homes Through ML-Driven Algorithms

This case study uses ML-driven algorithms to enable personalized environmental control in smart homes. Algorithms learn individual preferences and trends by analyzing input from numerous sensors and user choices. It changes ambient factors including temperature, lighting, and shade in real time to provide a personalized and comfortable living experience for each user. This personalized approach increases the user's experience and pleasure, resulting in a smart home that adjusts to the occupants' specific requirements and preferences [15]. Algorithms respond to changing user preferences and environmental circumstances through continuous data analysis and machine learning, allowing environmental controls to maintain personalized responsiveness over time [20]. To optimize ambient conditions, the system takes into account parameters such as time of day, occupancy, and even external elements such as weather forecasts. Users may also interact with the system using voice commands or the mobile phone app to change their settings and offer real-time feedback, resulting in a smooth and intuitive experience [32]. By leveraging ML-powered algorithms for personalized environmental control, smart homes can create harmonious and customized atmospheres, improving comfort, well-being, energy efficiency, and ultimately the way occupants live their lives. It can change the way you interact with and enjoy your space [22].

8 Analysis

The comparative analysis of the four case studies provides a detailed examination of the outcomes, highlighting specific technological implementations, results, and comparisons across different smart home scenarios.

8.1 Energy Management in Smart Homes with Machine Learning, IoT, and Cloud Computing

Technological Implementations: Smart plugs, lighting systems, appliances, and thermostats equipped with IoT sensors collect real-time data on occupancy, weather, and energy usage. This data is transmitted to the cloud for storage and analysis. ML algorithms analyze the data to identify patterns, predict energy usage trends, and suggest energy-saving recommendations.

Results: The implementation led to a significant reduction in energy waste and operational costs. Personalized recommendations helped homeowners optimize their energy consumption, contributing to sustainability efforts.

Comparison: This case study highlights the scalability and efficiency of cloud-based energy management systems compared to traditional methods. The use of ML algorithms for predictive analytics sets it apart, enabling more accurate and personalized energy optimization.

8.2 Intelligent Security Systems for Smart Homes Leveraging ML Algorithms

Technological Implementations: IoT devices such as security cameras, motion detectors, door/window sensors, and smart locks collect data on environmental conditions and security incidents. This data is sent to the cloud, where ML algorithms analyze it to detect patterns and anomalies.

Results: The system enhanced home security through real-time monitoring, intrusion detection, and remote access features. Homeowners received immediate alerts on potential security breaches, improving response times and reducing the risk of unauthorized access.

Comparison: The integration of IoT devices with cloud-based ML algorithms offers superior real-time monitoring and anomaly detection capabilities compared to standalone security systems. The ability to remotely access and control security devices provides an added layer of convenience and safety.

8.3 Predictive Maintenance of Smart Home Devices Using Cloud-Based Analytics

Technological Implementations: IoT devices continuously monitor the health and performance of various smart home appliances. Data is collected and analyzed in the cloud using ML models to predict potential failures and schedule maintenance.

Results: Predictive maintenance reduced downtime and improved the reliability and lifespan of smart home devices. Homeowners benefited from efficient maintenance planning and resource allocation, leading to cost savings.

Comparison: Compared to reactive maintenance approaches, predictive maintenance using cloud-based analytics offers proactive solutions, reducing unexpected failures and maintenance costs. The continuous learning capabilities of ML models enhance the accuracy and effectiveness of maintenance predictions over time.

8.4 Personalized Ambient Control in Smart Homes Through ML-Driven Algorithms

Technological Implementations: Sensors and user preferences data are analyzed by ML algorithms to adjust ambient conditions such as temperature, lighting, and shading in real-time. The system learns individual preferences and adapts accordingly.

Results: The personalized ambient control system significantly enhanced user comfort and satisfaction by creating tailored living environments. Energy efficiency was also improved as the system optimized settings based on occupancy and external factors.

Comparison: This approach stands out for its focus on personalization and user experience. Table 2 provides a comparative analysis of the case studies, highlighting key technologies, applications, and benefits. Unlike static control systems, ML-driven algorithms provide dynamic adjustments, ensuring that the living environment continually adapts to the occupants' needs and preferences.

These case studies demonstrate his various applications of IoT, ML and cloud computing in smart homes. Each case study covers a specific aspect, from energy management to safety, maintenance and personalized comfort. However, they are all aimed at improving functionality, efficiency and user experience in the smart home environment. The key findings are:

- The combination of IoT, machine learning, and cloud computing offers sophisticated smart home capabilities. Energy optimization, improved security, predictive maintenance, and individual environmental controls.

Table 2 Comparative analysis of case studies

S. No.	Case study	Focus	Key technologies	Applications	Benefits
1	Energy management	Optimizing energy consumption and efficiency	IoT devices, cloud computing, ML algorithms	Real-time energy monitoring, optimization, personalized recommendations	Reduced energy waste, cost savings, sustainability
2	Intelligent security systems	Enhancing security and monitoring capabilities	IoT devices, cloud computing, ML algorithms	Intrusion detection, anomaly detection, remote monitoring and control	Improved home security, real-time alerts, remote access
3	Predictive maintenance	Proactively maintaining smart home devices	Cloud-based analytics, ML models, IoT devices	Fault detection, equipment health monitoring, predictive maintenance planning	Increased device reliability, cost savings, efficient resource allocation
4	Personalized ambient control	Creating personalized and comfortable living environments	ML-driven algorithms, sensor data, user preferences	Adjusting temperature, lighting, and other ambient conditions based on individual preferences	Enhanced user experience, customized comfort, energy efficiency

- ML algorithms play a key role in analyzing data collected from sensors and devices, enabling intelligent decision-making, anomaly detection, and personalized experiences.
- Cloud-based infrastructure provides scalability, storage, and computing power to facilitate real-time data processing, analytics, and remote access to smart home systems.
- Personalization is a key aspect of case studies, aimed at creating customized experiences based on individual preferences and behavioral patterns.
- The use of these technologies provides benefits such as energy savings, improved security, enhanced user comfort, optimized maintenance, and ultimately a better overall smart home experience.

9 Challenges and Limitations Observed in the Case Studies

The case studies identified several challenges and limitations in the integration of IoT, ML, and cloud computing in smart homes. These challenges must be addressed to ensure a secure, efficient, and user-centric smart home ecosystem. Below are the challenges and limitations identified:

9.1 Privacy and Security

Challenges: Collecting and analyzing personal data in smart homes require robust privacy measures to protect user information. There is a significant risk of data breaches and unauthorized access.

Solutions: Implement advanced encryption techniques, ensuring data anonymization, enforcing strict data access controls, behavioral biometrics, regular security updates, and comprehensive privacy policies to safeguard data. Educate users on best practices for securing their smart home devices. Exploring the use of blockchain for decentralized and tamper-proof data storage. Conducting frequent security audits and updates to address emerging threats.

9.2 Compatibility and Interoperability:

Challenges: Diverse systems and devices often face compatibility and interoperability issues, impeding smooth data transmission and integration, which can affect overall functionality and performance.

Solutions: Promote the adoption of standard protocols and interfaces. Develop interoperable solutions that ensure seamless communication between different devices and systems.

Challenges: Smart home devices from different manufacturers often lack interoperability, leading to fragmented systems.

Solution: Adopting universal standards and protocols can facilitate better interoperability and create a cohesive smart home ecosystem.

9.3 Technical Expertise and Investment:

Challenges: Deploying and maintaining IoT devices, ML algorithms, and cloud infrastructure require significant technical expertise, initial investment, and ongoing management.

Solutions: Provide training and support for homeowners and technicians. Explore cost-effective solutions and subsidies to lower the entry barrier for smart home technologies.

9.4 Data Quality and Availability:

Challenges: ML algorithms depend on accurate and representative training data. Poor data quality and availability can hamper the effectiveness of these algorithms.

Solutions: Implement rigorous data collection and validation processes. Use diverse datasets to train ML models, ensuring they perform well across various scenarios.

9.5 Complexity and Diversity:

Challenges: The complexity and diversity of smart home environments create challenges in developing generalizable and scalable solutions.

Solutions: Focus on modular and flexible system designs that can adapt to different home environments. Encourage ongoing research and development to create scalable solutions.

Addressing these challenges involves:

- Enhancing privacy and security measures.
- Promoting standardization and interoperability.
- Improving data quality and availability.
- Increasing user awareness and education.

- Enhancing technical capabilities for seamless integration and performance optimization.

The case studies demonstrate the transformative potential of IoT, ML, and cloud computing in smart homes. Despite the challenges, these technologies offer significant benefits, including energy efficiency, improved security, streamlined maintenance, and personalized comfort. Maintaining high standards of data privacy, promoting wider adoption, and continuous innovation are crucial for maximizing the potential of smart home technologies.

10 Opportunities for Further Research and Development

One of the most important future trends in smart homes is the integration of new technologies. As technology continues to evolve, new advancements such as 5G connectivity, augmented reality (AR), edge computing, and virtual reality (VR) could further enhance smart home capabilities. These technologies enable faster, more reliable connectivity, distributed computing at the edge, immersive user experiences, and improved interactions with smart home devices. The integration of these technologies paves the way for a smarter and more immersive smart home environment.

Privacy and security in the smart home remain a challenge. The quantity of private and sensitive data being gathered and transmitted is growing with the number of connected devices. In order to safeguard privacy and stop unwanted access to or exploitation of personal data, it is crucial to have strong security measures, data encryption, and user permission procedures. Additionally, fixing security vulnerabilities and implementing standardized security protocols across all smart home devices and platforms is critical to maintaining user trust.

Standardization and interoperability are necessary for smooth communication and integration across different smart home platforms and gadgets. The lack of common standards and protocols currently exists can create compatibility issues and affect the interoperability of devices from different manufacturers. Adoption of standardized communication protocols and data formats is critical to enabling real interoperability and easing the integration of diverse smart home systems.

Widespread adoption of smart home devices depends on improving their usability and user experience. The complexity of device setup, configuration, and user interfaces can be a barrier for homeowners who find smart home technology difficult to utilize and realize its full potential. Future advancements should prioritize user-friendly interfaces, reduced setup processes, and user-centered design to make smart homes more accessible, user-friendly, and inclusive.

As the demand for smart homes develops, it becomes increasingly necessary to ensure energy efficiency and sustainability. You can optimize your energy use, lower your carbon footprint, and contribute to a more sustainable, greener future by combining energy management systems, renewable energy sources, and sophisticated analytics. Smart home solutions that prioritize energy efficiency, enable demand response programs, and encourage green practices will be critical in solving environmental challenges.

As smart homes become more powerful, ethical issues become increasingly significant. We must guarantee that technology is used responsibly and ethically, including clear and transparent data collecting practices, unbiased algorithms, and fair decision-making procedures. Furthermore, addressing concerns such as algorithmic bias, privacy breaches, and possible data breaches should be prioritized in smart home development.

As smart homes become more powerful, ethical issues become increasingly significant. We must guarantee that technology is used responsibly and ethically, including clear and transparent data collecting practices, unbiased algorithms, and fair decision-making procedures. Furthermore, addressing concerns such as algorithmic bias, privacy breaches, and possible data breaches should be prioritized in smart home development.

11 Summary and Recommendations

The potential of IoT, machine learning, and cloud computing in smart homes is immense, offering significant benefits in terms of energy efficiency, security, maintenance, and user comfort. However, to fully realize these benefits, it is essential to address the aforementioned challenges and limitations. Researchers should focus on developing robust security protocols, creating energy-efficient algorithms, and enhancing interoperability standards. Additionally, more user-friendly designs and ethical frameworks are necessary to ensure these technologies are accessible and beneficial to all.

The chart given in Fig. 8 compares the current implementation level with future research opportunities across various critical areas in smart home technologies:

Privacy and Security: Current implementation (40%), Future opportunities (80%)
Standardization and Interoperability: Current implementation (50%), Future opportunities (70%)
Usability and User Experience: Current implementation (60%), Future opportunities (75%)
Energy Efficiency and Sustainability: Current implementation (65%), Future opportunities (85%)
Ethical Considerations: Current implementation (55%), Future opportunities (90%).

While IoT, machine learning, and cloud computing have already begun to transform smart homes, there is still significant room for growth and improvement. By addressing the current challenges and continuing to innovate, we can develop smart

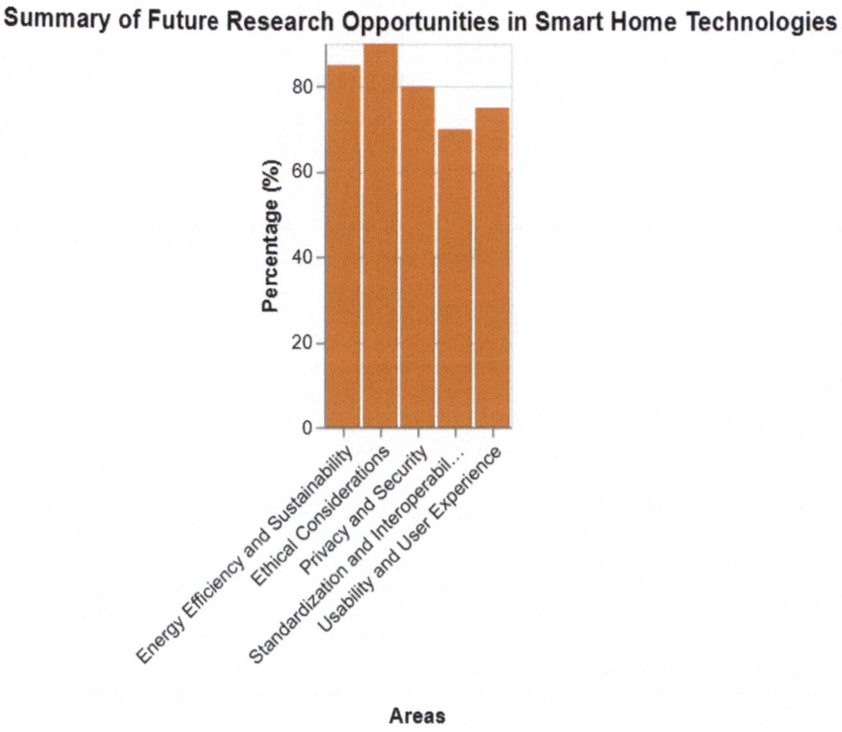

Fig. 8 Summary of future research opportunities

home technologies that are not only efficient and secure but also user-centric and ethically sound.

12 Conclusion

The study provides a comprehensive evaluation of the integration of IoT, ML, and cloud computing in smart homes, highlighting significant benefits such as energy efficiency, improved security, streamlined maintenance, and personalized comfort. The scientific contribution of this study includes the identification of current technological limitations and the proposal of innovative solutions to address these challenges. Future research should focus on developing robust security protocols, creating energy-efficient algorithms, and enhancing interoperability standards. Researchers are encouraged to explore these areas to advance the field and realize the full potential of smart home technologies (Table 3).

Table 3 Table of abbreviations

Abbreviation	Expansion
IoT	Internet of Things
ML	Machine Learning
AR	Augmented Reality
VR	Virtual Reality
HVAC	Heating, Ventilation, and Air Conditioning
AI	Artificial Intelligence
Wi-Fi	Wireless Fidelity

References

1. Botta A, De Donato W, Persico V, Pescapé A (2016) Integration of cloud computing and Internet of Things: a survey. Futur Gener Comput Syst 56:684–700
2. Singh P, Das L, Jha R, Kumar A, Rani S (2023) Revolutionizing sales and marketing: the power of AI in action. In: Security and risk analysis for intelligent cloud computing: methods, applications, and preventions, 1st edn. CRC Press, pp 297–315
3. Rani S, Tripathi K, Kumar A (2023) Machine learning aided malware detection for secure and smart manufacturing: a comprehensive analysis of the state of the art. Int J Interact Des Manuf (IJIDeM). https://doi.org/10.1007/s12008-023-01578-0
4. Phogat M, Kumar A, Nandal D, Shokhanda J (2021) A novel automating irrigation techniques based on artificial neural network and fuzzy logic. J Phys Conf Ser 1950(1):012088. https://doi.org/10.1088/1742-6596/1950/1/012088
5. Mohammadi M, Jamshidi S, Rezvanian A, Gheisari M, Kumar A (2024) Advanced fusion of MTM-LSTM and MLP models for time series forecasting: an application for forecasting the solar radiation. Measur Sens 33:101179
6. Naveena K, Krishnamoorthy M, Karuppiah N, Gouda PK, Hariharan S, Saravanan K, Kumar A (2024) Elevating sustainability with a multi-renewable hydrogen generation system empowered by machine learning and multi-objective optimization. Measur Sens 33:101192
7. Dalal S, Kumar A, Lilhore UK, Dahiya N, Jaglan V, Rani U (2024) Optimizing cloud service provider selection with firefly-guided fuzzy decision support system for smart cities. Measur Sens 35:101294. https://doi.org/10.1016/j.measen.2024.101294
8. Kulshrestha R, Ramani P, Thakur P, Kumar A, Dogra N, Murthy KVSR, Nandan D (2024) An energy-efficient clustering approach for wireless sensor networks to reduce hot-spot effect and idle listening energy consumption. Math Model Eng Probl 11(7):1729–1742. https://doi.org/10.18280/mmep.110704
9. Kaur R, Kaur A (2020) Cloud computing and machine learning: a comprehensive review. J Ambient Intell Humaniz Comput 11(9):3837–3864
10. Malche T, Maheshwary P (2017) Internet of Things (IoT) for building smart home system. In: International conference on I-SMAC (IoT in social, mobile, analytics and cloud) (I-SMAC). IEEE, pp 65–70
11. Ghayvat H, Liu J, Babu A, Alahi EE, Gui X, Mukhopadhyay SC (2015) Internet of Things for smart homes and buildings: opportunities and challenges. J Telecommun Digital Econ 3(4):33–47
12. Saleem AA, Hassan MM, Ali IA (2022) Smart homes powered by machine learning: a review. In: International conference on computer science and software engineering (CSASE), pp 355–361
13. Cook DJ, Crandall AS, Thomas BL, Krishnan NC (2013) CASAS: a smart home in a box. Comput 46(7):62–69

14. Lakhwani K, Gianey H, Agarwal N, Gupta S (2019) Development of IoT for smart agriculture a review. In: Emerging trends in expert applications and security: proceedings of ICETEAS 2018. Springer, pp 425–432
15. Priyadarshini I, Sahu S, Kumar R, Taniar D (2022) A machine-learning ensemble model for predicting energy consumption in smart homes. Internet Things 20(1):100636
16. Xiao G (2021) Machine learning in smart home energy monitoring system. In: IOP conference series: earth and environmental science, vol 769, no 4, p 042035
17. Sinha S, Raj T, Kumar S (2023) Smart home automation using IoT with cloud computing. Retrieved from ResearchGate
18. Chataut R, Phoummalayvane A, Akl R (2023) Unleashing the power of IoT: a comprehensive review of IoT applications and future prospects in healthcare, agriculture, smart homes, smart cities, and industry 4.0. Sensors 23(16):7194
19. Magara T, Zhou Y (2024) Internet of Things (IoT) of smart homes: privacy and security. J Electr Comput Eng 2024(1):7716956
20. Padmanaban S, Nasab MA, Shiri ME, Javadi HHS, Nasab MA, Zand M, Samavat T (2023) The role of internet of things in smart homes. In: Artificial intelligence-based smart power systems, pp 259–271
21. Xu Y, Khorasgani H, Ma J, A survey on data privacy in smart homes. J Ambient Intell Humaniz Comput 10(9):3837–3860
22. Majeed A, Raza B, Ali R, Ahmad R (2020) Energy management in smart homes: a comprehensive review. Energies 13(7):1795
23. Lakhwani K, Sharma G, Sandhu R, Nagwani NK, Bhargava S, Arya V, Almomani A (2023) Adaptive and convex optimization-inspired workflow scheduling for cloud environment. Int J Cloud Appl Comput (IJCAC) 13(1):1–25
24. Soliman M, Abiodun T, Hamouda T, Zhou J, Lung C-H (2013) Smart home: integrating Internet of Things with web services and cloud computing. In: 5th International conference on cloud computing technology and science, 2nd edn, pp 317–320
25. Machorro-Cano I, Alor-Hernández G, Paredes-Valverde MA, Rodríguez-Mazahua L, Sánchez-Cervantes JL, Olmedo-Aguirre JO (2020) HEMS-IoT: a big data and machine learning-based smart home system for energy saving. Energies 13(5):1097
26. Djumanazarov O, Väänänen A, Haataja K, Toivanen P (2022) An overview of IoT-based architecture model for smart home systems. In: Abraham A, Gandhi N, Hanne T, Hong TP, Nogueira Rios T, Ding W (eds) International conference on intelligent systems design and applications, ISDA 2021. Lecture notes in networks and systems, vol 418. Springer, Cham, pp 696–706
27. Sharma VK, Mohapatra SK, Shitharth S, Yonbawi S, Yafoz A, Alahmari S (2022) An optimization-based machine learning technique for smart home security using 5G. Comput Electr Eng 104:108434
28. Yang S (2022) Application of cloud computing technology in smart home modeling design. In: 3rd International conference on artificial intelligence and advanced manufacture (AIAM2021). Association for Computing Machinery, pp 989–992
29. Peng Y, Peng J, Li J, Yu L (2019) Smart home system based on deep learning algorithm. J Phys: Conf Ser 1187:032086
30. Shafi MKI, Sultan MR, Rahman SMM, Hoque MM (2021) IoT based smart home: a machine learning approach. In: 24th International conference on computer and information technology (ICCIT), pp 1–6
31. Kar HA, Rather GM (2019) A Survey on edge-based Internet-of-Things. Int J of Comput Netw Appl (IJCNA) 6(6):100–109
32. Lakhwani K, Gianey HK, Wireko JK, Hiran KK (2020) Internet of Things (IoT): principles, paradigms and applications of IoT, 1st edn. Bpb Publication, New Delhi
33. Hiran KK, Jain RK, Lakhwani K, Doshi R (2021) Machine learning: master supervised and unsupervised learning algorithms with real examples (English Edition). BPB Publications, New Delhi

Comparative Analysis of Malware Detection Techniques and Machine-Learning Algorithms Used for Security Testing of Android Applications

Pawan Kumar and Sukhdip Singh

Abstract Android application security testing comprises testing a mobile app through attacks that a malicious user would try. Android application security testing is a new field that is currently developing. The purpose of conducting this review is to briefly summaries the role of malware detection techniques and machine learning in security testing used for Android applications. The objective of this article is to gain some level of understanding of malware detection techniques and to analyze the machine learning algorithms used for security testing of Android applications. The findings of this study are intended to provide direction to security practitioners, researchers, and application developers in selecting appropriate methods and algorithms for protecting Android applications. This research supports ongoing initiatives to strengthen the vast Android app ecosystem and protect user data from nefarious actors by emphasizing the advantages and drawbacks of different techniques.

Keywords Android applications · Security testing · Malware detection · Machine-learning algorithms · Comparative analysis · Security · Privacy

1 Introduction

Mobile phones are a primary part of our daily life. It plays an important role in our daily activities that includes making calls, sending text messages, capturing photos, doing financial transactions, and playing games. There are millions of applications available on play stores. To be famous among users, these applications should be easy, safe, and secure. Regretfully, hackers are particularly interested in privacy-related data. As a result, hackers are developing new mechanism and methods to steal data from devices.

P. Kumar (✉) · S. Singh
Department of Computer Science and Engineering, Deenbandu Chhotu Ram University of Science and Technology, Murthal, India
e-mail: pawankumar.schcse@dcrustm.org

A. Kumar et al. (eds.), *Trends in Mechatronics Systems*, Emerging Trends in Mechatronics, https://doi.org/10.1007/978-981-97-9108-8_9

Security testing of Android application is the latest and emerging research area that requires knowledge and skills. There are many methods for security testing these event-based applications, but malware analysis is the latest emerging and efficient method. Research is going on to develop more efficient security testing methods using various machine learning algorithms.

Security testing of traditional java applications is much easier as they have one entry point from a program launcher, and then run as a unique monolithic process. On the contrary Android operating system has a much more complex architecture. So, it is very difficult to check all functionality of Android applications.

Security testing of Android applications is little difficult because programmers must consider the varieties of configuration in hardware. In general, security testing is increasingly hard when mixed with malware analysis and machine learning. Security testing of Android apps requires not only knowledge of programming languages (like Java, Python) for analysis of Android applications or their apk files but also machine learning algorithms.

Many users of Android smartphone know about the **Camscanner** app which converts photos of the physical documents into PDF and image files. Kaspersky found a malicious **Trojan** Dropper module in the advertising library of its 5.11.7 version. As a result of it, Google took the app down. This incident left a scare on Camscanner reputation and distrust among its users. This happened due to the lack of security testing methods. This incident highlights the significance of Android app security testing in the ever-growing world of technology [1].

Malware analysis is the process of examining harmful software to learn about its features, actions, and motivations. It plays a crucial role in security testing. In order to recognize and categorize various forms of malware, including viruses, worms, Trojan horses, and ransomware, malware analysis is used. Having a thorough understanding of malware's traits aids security experts in creating defenses. Security systems like antivirus software use signatures or patterns that analysts establish in order to detect and stop malware attacks. Based on the distinctive characteristics of particular malware strains, these signatures. Malware analysis entails running malware samples in supervised settings (like sandboxes) to watch how they behave.

Security specialists can better analyze the malware's actions and interactions with the system thanks to this. The knowledge gathered from malware analysis is applied to the development of tactics for reducing malware threats, including the development of patches, the updating of security guidelines, and the strengthening of network defenses.

By automating and improving many components of the process, machine learning, a subset of artificial intelligence, contributes significantly to security testing. To make forecasts and judgments about security concerns, it uses data-driven models.

The objective of the research includes following:

1. Current status of Android operating system.
2. Role of malware analysis in security testing of Android applications.
3. Role of machine learning in security testing of Android applications.
4. Comparative analysis of latest articles in security testing of Android applications.

The structure of this article is as follows: Sect. 2 provides an overview of the Android operating system, while Sect. 3 delves into the current status of the Android OS. Section 4 focuses on security testing in Android applications, and Sect. 5 is dedicated to malware analysis. In Sect. 6, we explore the role of machine learning, and Sect. 7 presents the findings and results. The final section, Sect. 8, offers a conclusion.

This research work identifies that in the present literature surveys the hybrid analysis has not been included. This research work will include the hybrid analysis along with static analysis and dynamic analysis and it will be very helpful for further investigation in the Android application security testing area.

2 Android Operating System

The foundation of Android software stack as shown Fig. 1 consists of Linux kernel to handle threading and low-level memory management like issues. Linux kernel provides security features and is also used to develop hardware drivers. HAL (Hardware Abstraction Layer) includes library modules that are used for implementing the interface for hardware components. ART (Android Run Time) is used to run many VM's on low memory mobile devices by using DEX files. It includes JIT, garbage collector, DEX machine code and debugger. Mainly ART and HAL are developed using the native libraries in C and C++.

The feature set of the Android operating system includes management of views, resources, activities, notifications, and content providers. All components of feature set are developed using java API framework. System Applications includes an email, a messaging app, a calendar, a browser, and a contact app. These applications are also known as core apps.

3 Status of Android Operating System

In this modern era, the usage of smart phones is increasing day by day. The usage is increasing for day-to-day activities (like calling a friend, photo capturing, using internet, making bill payment, etc.) are done by using mobile phones. Google collaborates with Smartphone makers, software companies, and hardware companies by creating an OHA (Open Handset Alliance) to compete with the Apple iOS mobile operating system. OHA established Android as the mobile platform of choice that allows an open-source license to mobile phone manufacturers. As per report74% of mobile users are using Android mobile phones [2].

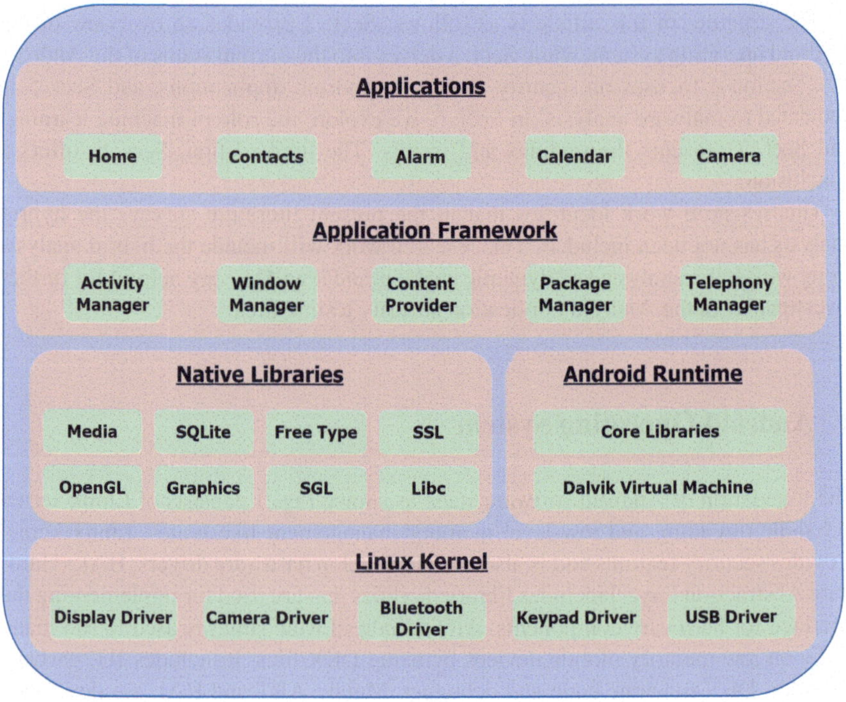

Fig. 1 Android software stack

4 Security Testing in Android Application

In this modern era, the usage of smart phones is increasing for day-to-day activities (like calling a friend, making bill payment, etc.) are performed by using Android applications. These applications should be user friendly, secure, and safe for usage. To do this, the Android application testing is a trending field in these days. UI testing, security testing and functional testing are the major testing performed on Android application [3–5]. Testing of Android application is a difficult task due to diversity in hardware configurations and Android have a large number of versions [6, 7].

The objectives of Android app security testing are to identify the purpose for which the testing has been conducted. In general terms, testing is the process of finding out how well something works. Software testing is the process used to check whether objectives are being met. For example, Users of software product sometimes test their objective.

Android application security testing is a process by which Android application developed for smartphones is tested based on signatures and behavior of the application. Security testing of Android application is an important part of the testing process. Android application security testing includes verification of the code without executing it or verify the behavior of the application on device/emulator. It verifies the usability, correctness, and functional behavior of Android application before releasing it publicly.

Android application can be tested using automated testing or manual testing. Although Android is an open-source operating system but Android applications are both open source and closed source. Open-source Android applications are easy to test as the source code is easy to access. F-Droid is the best source for the open-source Android applications. The source code of many open-source Android applications is available on GitHub while the Google play store is the main source of closed-source applications. The source code of these applications is not directly available, so it is very difficult to identify malicious applications.

Many techniques were proposed in present years for identifying the defects, the bugs, and the errors in the applications to enhance the quality of these applications. Malware analysis is a major activity used for defect identification [8].

Security is the major requirements of an app. However, in financial and healthcare application this becomes the primary requirement. Testing of encryption/decryption mechanism and access to stored data is checked in security testing.

5 Malware Analysis

The usage of mobile devices has grown significantly in the past decades, and with that, so has the proliferation of Android malware [9]. With increased mobile phone use comes an increase in Android malware. Because there are so many people using the Android operating system, it is quite simple and easy for malware to flourish and propagate.

Hu et al. [10] developed a system AppDoctor, for efficiently testing Android apps and for diagnosing the resultant bug reports. AppDoctor quickly detect potential bugs using approximate execution and also use action slicing to speed up bug diagnosis. AppDoctor saves manual inspection effort by automatic verification of bugs from the reports.

According to recent research, several experts have created their own approaches for identifying malware and malicious applications on Android. The following three categories serve as a general breakdown of malware detection approaches.

5.1 Static Analysis

Static analysis refers to the use of analytical techniques to the study of any application in which there is no code execution. The process used to examine the application package is called static analysis as shown in Fig. 2. The .apk file of the relevant application is analyzed as part of this.

The process of downloading an application begins in the Google play store. Source code file class.dex and AndroidManifest.xml file are the only two that are examined

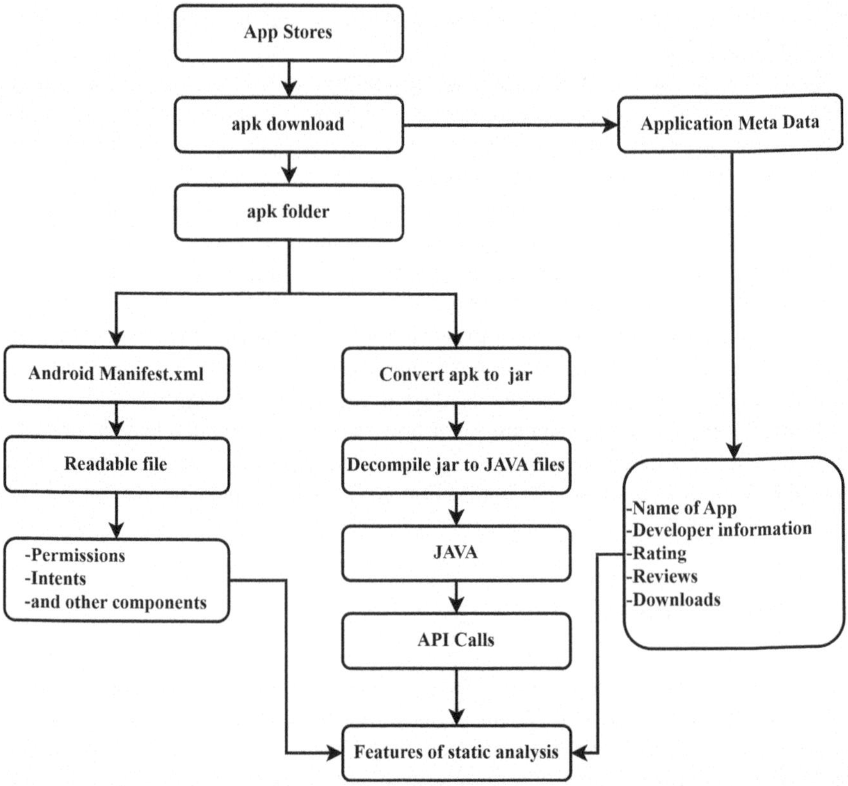

Fig. 2 Static analysis process

throughout the static analysis process of application folders (.apk). A package's permissions, hardware, software, and other information are all mentioned in the Android manifest file. Source code file, on the other hand, contains intents, system events, API call information, and other details [11].

Recent studies have shown several features that are employed in the static analysis of programs to find malicious ones. The procedure of detection makes use of the permissions used by both malicious and benign applications.

It can be utilized independently or in conjunction with one or more other features. Android application's apk file contains manifest file is where permissions are extracted. Permissions are important because they provide the details of the components and resources needed for execution of an application on a device. So, based on the permissions requested, researchers can analyze the behavior of the application. Other components used for analyzing the behavior of applications are API calls, intents, and Opcode.

5.2 *Dynamic Analysis*

Dynamic analysis is used to investigate applications by executing the code and observing the behavior of the application as depicted in Fig. 3. For dynamic analysis, two methodologies are used: in-box and out-of-box analysis [12]. Both methodologies require a separate environment for execution of the applications and observe the application behavior on real device. Dynamic analysis technique is more complicated than static analysis technique and requires more skill sets and resources to observe and draw conclusions.

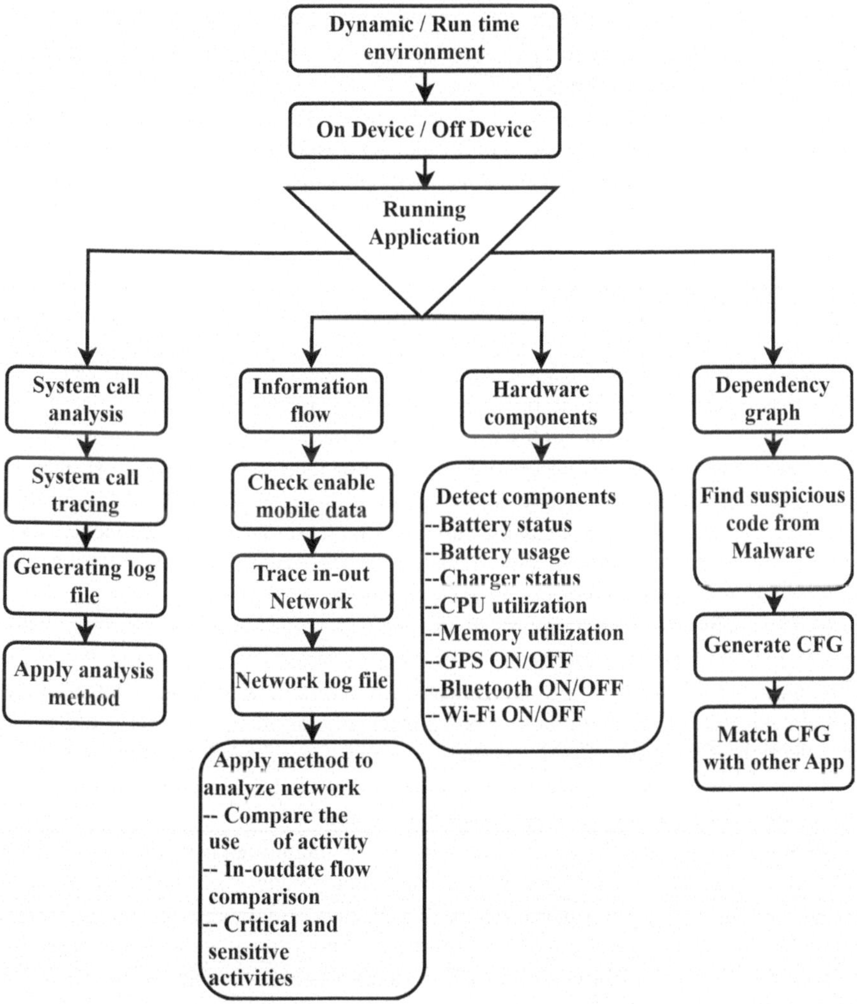

Fig. 3 Dynamic analysis process

Su et al. [13] used a guided approach Stoat, which executed stochastic testing on applications under test. Stoat operate in two steps, the first step includes Android app as input to the dynamic analysis that comprises weighted user interface exploration strategy as well as static analysis for regenerating the stochastic model for the GUI of Android app. The second step uses Gibbs sampling for refining the model and is also used for guiding test case generation from the model mutation for achieving high coverage, releasing multiple sequences. Testing process injects random system level events for increasing testing efficiency.

The activity of the mobile application is evaluated in dynamic analysis for malware analysis. There are two methods for analyzing a mobile application: in-box and out-of-box. The mobile application is executed and evaluated for the behavioral examination, such as information flow, system calls, hardware component consumption, and so on. These are the primary features that can be processed for the application's behavior study during run time.

System calls, information flow, function call monitoring, dependency graphs, hardware component analysis, used API and permissions, code injection, runtime user interface and dynamic code loading are the features used for dynamic analysis.

5.3 Hybrid Analysis

The integration of multiple static analysis and dynamic analysis approaches is known as hybrid analysis. It involves combining and comparing the results of different malware detection techniques and machine learning algorithms to obtain a more comprehensive and accurate assessment of their performance [14]. It may entail running Android applications through multiple detection methods, including signature-based, behavioral analysis, and machine learning-based approaches. The results obtained from these diverse methods are then systematically compared to identify strengths and weaknesses in each technique.

By using a hybrid analysis approach, the research seeks to provide a more holistic understanding of the capabilities and limitations of various security testing methods. This approach allows for a more well-rounded assessment of how different techniques and algorithms complement or enhance one another, ultimately aiding in the selection of the most effective strategies for securing Android applications. The hybrid analysis approach can provide a more robust evaluation and contribute to a more comprehensive and valuable research outcome.

Most researchers employed this combination by including permissions and APIs in the static segment, and system calls are used in the dynamic segment. To facilitate malware analysis by reintroducing the traditional analyze-then-execute methodology, which has the advantage of placing harmful code under the analyst's control before it executes. The control-flow and data-flow analyses that serve as the interface for the analyst to interpret and instrument the code are built and maintained using static and dynamic analysis approaches.

6 Machine Learning

Machine learning is a branch of artificial intelligence that allows computers to learn without being explicitly programmed. Machine learning [14–16] is used for security testing of Android applications. These techniques demand more contribution from research community. Android applications are the combination of multiple components like activities, broadcast receivers, content providers, fragments, and intents. Machine-learning techniques are employed to enhance the analysis and assessment of various malware detection methods and their effectiveness in securing Android applications. Here are some specific roles of machine learning in this comparative analysis:

Automated Analysis: Machine-learning algorithms can be utilized to automatically analyze a large dataset of Android applications and the malware detection techniques applied to them. This automation streamlines the analysis process, saving time, and resources.
Feature Extraction: Machine learning can extract relevant features from Android applications and their corresponding security testing results. These features may include code attributes, behavioral patterns, and indicators of malware presence or absence.
Pattern Recognition: Machine-learning algorithms can recognize patterns in the data, helping to identify trends and correlations between the choice of malware detection techniques, the characteristics of Android applications, and the detection outcomes.
Performance Assessment: Machine-learning models can be used to quantitatively assess the performance of different malware detection techniques. They can measure accuracy, precision, recall, and other relevant metrics to determine how well these techniques work in practice.
Recommendation Systems: Machine learning can aid in the development of recommendation systems that suggest the most appropriate malware detection techniques for specific types of Android applications or threat scenarios, based on historical data and analysis.
Adaptability and Scalability: Machine-learning models can adapt to changing threat landscapes and evolving Android security paradigms. This adaptability ensures that the analysis remains relevant over time.
Insight Generation: By analyzing the data and the results of security testing, machine learning can generate insights and identify the strengths and weaknesses of different malware detection techniques and algorithms.
Reduction of Bias: Machine learning can help minimize human bias in the analysis process, ensuring a more objective and data-driven assessment of security testing methods.
Predictive Analysis: Machine-learning models can make predictions about the future performance of malware detection techniques, helping security professionals anticipate potential vulnerabilities and threats.

Optimization: Machine learning can assist in optimizing the use of resources by identifying the most cost-effective and efficient malware detection techniques for specific applications and contexts.

In summary, machine learning plays a central role in the comparative analysis of malware detection techniques and machine learning algorithms used for security testing of Android applications by automating, enhancing, and providing data-driven insights into the effectiveness and efficiency of these techniques. It facilitates a more systematic and comprehensive evaluation of security measures in the Android ecosystem, ultimately contributing to the development of stronger security practices.

7 Results and Discussions

A comprehensive overview of various research studies focused on the detection of malware in Android applications. These studies employ a range of detection techniques and machine learning algorithms, with the associated accuracy rates, in the context of security testing for Android applications.

Su et al. [13] uses static analysis with features such as permissions, intents, and API calls, combined with kNN and K-means machine learning algorithms, achieving an accuracy of 97.87%. Arp et al. [14] also employs static analysis using SVM with features like permissions, network access, and API calls, achieving a 94% accuracy rate. Yerima et al. [15] relies on static analysis and ensemble learning, utilizing features like permissions, intents, and API calls, resulting in a high accuracy rate of 98%. Varsha et al. [16] conducts static analysis with features like permissions, filtered intents, opcodes, and strings. They use SVM, random forests, and rotation forests, achieving an accuracy rate of 97.2%.

Sharma and Dash [17] use static analysis with features such as API calls and permissions, combining Naive Bayes and kNN algorithms to attain an accuracy rate of 95.1%. Wang et al. [18] also employs static analysis, focusing on permissions, and uses SVM, random forest, and decision trees, resulting in an accuracy rate of 94.62%. DroidDolphin [19] opts for dynamic analysis of log files using big data analysis and achieves an accuracy rate of 86.1%.

Crowdroid [20] employs dynamic analysis on all generated log files using k-means, with an accuracy rate of 85%. DroidWard [21] conducts dynamic analysis with features like API calls and packet size. They utilize SVM, DT, and RF to reach an impressive accuracy rate of 98.54%. Fan et al. [22] uses static analysis with features like API calls and permissions. They employ kNN, Random Forest, Decision Tree, and PART algorithms, achieving an accuracy rate of 96%. Cen et al. [23] relies on static analysis with features like permissions and API calls, using Regularized Logistic Regression to achieve an accuracy rate of 96.7%.

Kang et al. [24] conducts static analysis with n-gram opcode features and employs NB, SVM, RF, and PART algorithms to achieve an accuracy rate of 96.88%. Elsersy et al. [25] employs a hybrid analysis approach, using features such as API calls and permissions, and applies LASSO regression to achieve a high accuracy rate of 98.16%. Kabakus et al. [26] relies on static analysis with features like intents, API calls, and permissions, using a CNN algorithm to achieve an accuracy rate of 98.36%. DeepAMD [27] uses dynamic analysis with features such as API calls and all generated log files. They employ J48, NB, and MLP algorithms, achieving an accuracy rate of 93.4%.

Daoudi et al. [28] uses a hybrid analysis approach with features including API calls, permissions, and network addresses. They apply SVM and achieve an accuracy rate of 97.9%. OverlayChecker [29] conducts dynamic analysis using overlay features with NB, LR, and RF algorithms, achieving an accuracy rate of 96%. Wu et al. [30] opts for static analysis with API features and employs the k-nearest neighbor algorithm, achieving an accuracy rate of 97.66%. Taha et al. [31] uses static analysis with Android permissions and applies Adaptive and Choquet fuzzy algorithms, achieving an accuracy rate of 95.08%.

MalPat [32] conducts static analysis with features like intents, API calls, and permissions, using Random Forest, resulting in an accuracy rate of 96.7%. MUDFLOW [33] relies on static analysis with features such as API calls and permissions, using SVM and achieving an accuracy rate of 96.4%. DroidAPIMiner [34] uses static analysis with Android permissions and KNN, resulting in an accuracy rate of 96%. DroidSIFT [35] conducts static analysis with features like API calls and system calls, employing RF and achieving an accuracy rate of 97.23%. DroidLegacy [36] uses static analysis with features like API calls, system calls, and resource usage, and applies SVM to achieve an accuracy rate of 95.68%.

Overall, the Table 1 provides insights into the diverse techniques and algorithms used for malware detection in Android applications. The varying accuracies of these methods highlight the effectiveness of different approaches, and the results can guide researchers and practitioners in choosing appropriate strategies for security testing in the Android app ecosystem. The studies demonstrate the importance of considering both static and dynamic analysis as well as hybrid approaches to achieve robust and accurate security testing results (Fig. 4).

8 Conclusion

The aim of the study is to analyze various Android app security testing techniques. This study includes security and defect identification for Android application testing. Defect Identification has been the mostly used test objective. In this research, all articles have used defect identification for test objective identification. Defect identification included finding an error/bug/defect within source code or functionality of the Android application.

The size of dataset should be large as it would augment the accuracy of the testing process. Android app testing of open-source application gets more attentiveness from research community.

Table 1 Comparative analysis of recent articles

S. No.	Research	Detection technique	Features used	ML algorithm	Accuracy (%)
1	DroidMat [17]	Static analysis	Permissions, intents, and API Calls	kNN, K-means	97.87
2	Arp et al. [18]	Static analysis	Permissions, network access and API calls,	SVM	94
3	Yerima et al. [19]	Static analysis	Permissions, intents, and API Calls	Ensemble learning	98
4	Varsha et al. [20]	Static analysis	Permission, filtered intents, opcodes and strings	SVM, random forests, and rotation forests	97.2
5	Sharma and Dash [21]	Static analysis	API calls and permissions	Naive Bayes and kNN	95.1
6	Wang et al. [22]	Static analysis	Permissions	SVM, random forest and decision trees	94.62
7	DroidDolphin [23]	Dynamic analysis	Log files	Big data Analysis	86.1
8	Crowdroid [24]	Dynamic analysis	All generated log files	k-means	85
9	DroidWard [25]	Dynamic analysis	API Calls, Packet size	SVM, DT, RF	98.54
10	Fan et al. [26]	Static analysis	API calls and permissions	kNN, Random forest, decision tree, and PART	96
11	Cen et al. [27]	Static analysis	Permissions and API calls	Regularized logistic regression	96.7
12	Kang et al. [28]	Static analysis	n-gram opcode	NB, SVM RF, and PART	96.88
13	Elsersy et al. [29]	Hybrid analysis	API calls, permissions	LASSO regression	98.16
14	Kabakus et al. [30]	Static analysis	Intents, API calls and permissions	CNN	98.36
15	DeepAMD [31]	Dynamic analysis	API calls and all generated log files	J48, NB and MLP	93.4
16	Daoudi et al. [32]	Hybrid analysis	API calls, permissions, and network addresses	SVM	97.9
17	OverlayChecker [33]	Dynamic analysis	Overlay features	NB, LR, and RF	96
18	Wu et al. [34]	Static analysis	APIs	k-nearest neighbor	97.66
19	Taha et al. [35]	Static analysis	Android permissions	Adaptive and Choquet fuzzy	95.08

(continued)

Table 1 (continued)

S. No.	Research	Detection technique	Features used	ML algorithm	Accuracy (%)
20	MalPat [36]	Static analysis	Intents, API calls and permissions	RF	96.7
21	MUDFLOW [37]	Static analysis	API calls and permissions	SVM	96.4
22	DroidAPIMiner [38]	Static analysis	Android permissions	KNN	96
23	DroidSIFT [39]	Static analysis	API calls and system calls	RF	97.23
24	DroidLegacy [40]	Static analysis	API calls, system calls, and resource usage	SVM	95.68

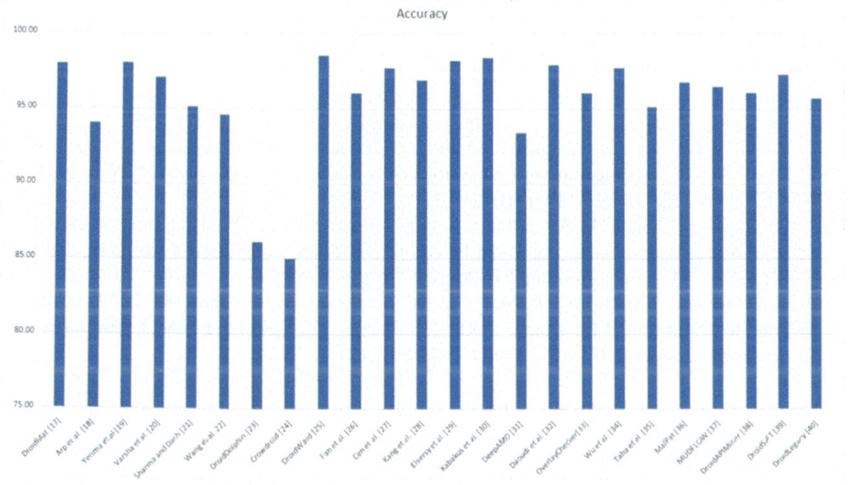

Fig. 4 Comparison of recent articles based on accuracy

References

1. CamScanner is a malicious Android app with more than 100 million downloads in Google Play. Kaspersky official blog
2. Google Play Store: number of apps. Statista
3. Amalfitano D, Fasolino AR, Tramontana P, De Carmine S, Imparato G (2012) A toolset for GUI testing of android applications. In: 2012 28th IEEE international conference on software maintenance (ICSM). IEEE, pp 650–653. https://doi.org/10.1109/ICSM.2012.6405345
4. Yeh CC, Lu HL, Chen CY, Khor KK, Huang SK (2014) CRAXDroid: Automatic android system testing by selective symbolic execution. In: 2014 IEEE eighth international conference on software security and reliability-companion. IEEE, pp 140–148. https://doi.org/10.1109/SERE-C.2014.32

5. Lin Y-D, Chu ETH, Yu S-C, Lai Y-C (2014) Improving the accuracy of automated GUI testing for embedded systems. IEEE Softw 31(1):39–45. https://doi.org/10.1109/MS.2013.100
6. Grano G, Ciurumelea A, Panichella S, Palomba F, Gall HC (2018) Exploring the integration of user feedback in automated testing of android applications. In: 2018 IEEE 25th international conference on software analysis, evolution and reengineering (SANER). IEEE, pp 72–83. https://doi.org/10.1109/SANER.2018.8330198
7. Jiang B, Wu Y, Zhang Y, Zhang Z, Chan KW (2018) ReTestDroid: towards safer regression test selection for android application. In: 2018 IEEE 42nd Annual computer software and applications conference (COMPSAC). IEEE, pp 235–244. https://doi.org/10.1109/COMPSAC.2018.00037
8. Jensen CS, Prasad MR, Møller A (2013) Automated testing with targeted event sequence generation. In: Proceedings of the 2013 international symposium on software testing and analysis. ACM, New York, pp 67–77. https://doi.org/10.1145/2483760.2483777
9. Rani S, Tripathi K, Kumar A (2023) Machine learning aided malware detection for secure and smart manufacturing: a comprehensive analysis of the state of the art. Int J Interact Des Manuf. https://doi.org/10.1007/s12008-023-01578-0
10. Hu G, Yuan X, Tang Y, Yang J (2014) Efficiently, effectively detecting mobile app bugs with appDoctor. In: Proceedings of the 9th European conference on computer systems, EuroSys 2014. https://doi.org/10.1145/2592798.2592813
11. KumarP, Singh S, Suman (2024) An efficient security testing mechanism for android apps based on malware analysis and optimized XGBoost. Multimed Tools Appl. https://doi.org/10.1007/s11042-023-18066-y.
12. Sharma T, Rattan D (2021) Malicious application detection in android—a systematic literature review. Comput Sci Rev 40:100373. https://doi.org/10.1016/j.cosrev.2021.100373
13. Su T et al (2017) Guided, stochastic model-based GUI testing of android apps. In: Proceedings of the 2017 11th joint meeting on foundations of software engineering. ACM, New York, pp 245–256. https://doi.org/10.1145/3106237.3106298
14. Kumar P, Singh S (2024) An efficient security testing for android application based on behavior and activities using improved PCA and DNN-KNN classifier. Wirel Pers Commun 135(2):1285–1314. https://doi.org/10.1007/s11277-024-11128-w
15. Dalal S, Onyema EM, Kumar P, Maryann DC, Roselyn AO, Obichili MI (2022) A hybrid machine learning model for timely prediction of breast cancer. Int J Model Simulation Sci Comput https://doi.org/10.1142/S1793962323410234
16. Zaki J, Nayyar A, Dalal S, Ali ZH (2022) House price prediction using hedonic pricing model and machine learning techniques. Concurr Comput Pract Exp 34(27). https://doi.org/10.1002/cpe.7342
17. Wu D-J, Mao C-H, Wei T-E, Lee H-M, Wu K-P (2012) DroidMat: android malware detection through manifest and API calls tracing. In: 2012 Seventh Asia joint conference on information security. IEEE, pp 62–69. https://doi.org/10.1109/AsiaJCIS.2012.18
18. Arp D, Spreitzenbarth M, Hübner M, Gascon H, Rieck K (2014) Drebin: effective and explainable detection of android malware in your pocket. https://doi.org/10.14722/ndss.2014.23247
19. Yerima SY, Sezer S (2019) DroidFusion: a novel multilevel classifier fusion approach for android malware detection. IEEE Trans Cybern 49(2). https://doi.org/10.1109/TCYB.2017.2777960Y
20. Varsha MV, Vinod P, Dhanya KA (2017) Identification of malicious android app using manifest and opcode features. J Comput Virol Hacking Tech 13(2):125–138. https://doi.org/10.1007/S11416-016-0277-Z/TABLES/28
21. Sharma A, Dash SK (2014) Mining api calls and permissions for android malware detection. In: Lecture notes in computer science (including Subser Lect Notes Artif Intell Lect Notes Bioinformatics), vol 8813, pp 191–205. https://doi.org/10.1007/978-3-319-12280-9_13/COVER
22. Wang W, Wang X, Feng D, Liu J, Han Z, Zhang X (2014) Exploring permission-induced risk in android applications for malicious application detection. IEEE Trans Inf Forensics Secur 9(11):1869–1882. https://doi.org/10.1109/TIFS.2014.2353996

23. Wu W-C, Hung S-H (2014) DroidDolphin. In: Proceedings of the 2014 conference on research in adaptive and convergent systems. ACM, New York, pp 247–252. https://doi.org/10.1145/2663761.2664223
24. Burguera I, Zurutuza U, Nadjm-Tehrani S (2011) Crowdroid: behavior-based malware detection system for android categories and subject descriptors. In: Proceedings of the 1st ACM workshop on security and privacy in smartphones and mobile devices. ACM, New York, pp 15–26. https://doi.org/10.1145/2046614.2046619
25. Yang Y, Wei Z, Xu Y, He H, Wang W (2018) DroidWard: an effective dynamic analysis method for vetting android applications. Cluster Comput 21(1):265–275. https://doi.org/10.1007/s10586-016-0703-5
26. Fan M, Liu J, Wang W, Li H, Tian Z, Liu T (2017) DAPASA: detecting android piggybacked apps through sensitive subgraph analysis. IEEE Trans Inf Forensics Secur 12(8):1772–1785. https://doi.org/10.1109/TIFS.2017.2687880
27. Cen L, Gates CS, Si L, Li N (2015) A probabilistic discriminative model for android malware detection with decompiled source code. IEEE Trans Dependable Secur Comput 12(4):400–412. https://doi.org/10.1109/TDSC.2014.2355839
28. Kang B, Yerima SY, Sezer S, McLaughlin K (2016) N-gram opcode analysis for android malware detection. Int J Cyber Situational Aware 1(1):231–255. https://doi.org/10.48550/arxiv.1612.01445
29. Elsersy WF, Anuar NB, Razak MFA (2022) ROOTECTOR: robust android rooting detection framework using machine learning algorithms. Arab J Sci Eng 1–21. https://doi.org/10.1007/s13369-022-06949-5
30. Kabakus AT (2022) DroidMalwareDetector: a novel android malware detection framework based on convolutional neural network. Expert Syst Appl 206:117833. https://doi.org/10.1016/j.eswa.2022.117833
31. Imtiaz SI, ur Rehman S, Javed AR, Jalil Z, Liu X, Alnumay WS (2021) DeepAMD: detection and identification of android malware using high-efficient deep artificial neural network. Futur Gener Comput Syst 115:844–856. https://doi.org/10.1016/j.future.2020.10.008
32. Daoudi N, Allix K, Bissyandé TF, Klein J (2022) A deep dive inside DREBIN: an explorative analysis beyond android malware detection scores. ACM Trans Priv Secur 25(2):1–28. https://doi.org/10.1145/3503463
33. Yan Y et al (2019) Understanding and detecting overlay-based android malware at market scales. In: Proceedings of the 17th annual international conference on mobile systems, applications, and services. ACM, New York, pp 168–179. https://doi.org/10.1145/3307334.3326094.
34. Wu S, Wang P, Li X, Zhang Y (2016) Effective detection of android malware based on the usage of data flow APIs and machine learning. Inf Softw Technol 75:17–25. https://doi.org/10.1016/J.INFSOF.2016.03.004
35. Taha A, Barukab O, Malebary S (2021) Fuzzy integral-based multi-classifiers ensemble for android malware classification. Mathematics 9(22):2880. https://doi.org/10.3390/math9222880
36. Tao G, Zheng Z, Guo Z, Lyu MR (2018) MalPat: mining patterns of malicious and benign android apps via permission-related APIs. IEEE Trans Reliab 67(1). https://doi.org/10.1109/TR.2017.2778147
37. Avdiienko V et al (2015) Mining apps for abnormal usage of sensitive data. In: 2015 IEEE/ACM 37th IEEE international conference on software engineering. IEEE, pp 426–436. https://doi.org/10.1109/ICSE.2015.61
38. Aafer Y, Du W, Yin H (2013) DroidAPIMiner: mining API-level features for robust malware detection in android. In: Lecture notes of the institute for computer sciences, social-informatics and telecommunications engineering, vol 127. LNICST, pp 86–103. https://doi.org/10.1007/978-3-319-04283-1_6
39. Zhang M, Duan Y, Yin H, Zhao Z (2014) Semantics-aware android malware classification using weighted contextual API dependency graphs. In: Proceedings of the 2014 ACM SIGSAC conference on computer and communications security. ACM, Nov 2014, New York, pp 1105–1116. https://doi.org/10.1145/2660267.2660359

40. Deshotels L, Notani V, Lakhotia A (2014) DroidLegacy: automated familial classification of android malware. In: Proceedings of ACM SIGPLAN on program protection and reverse engineering workshop 2014. ACM, New York, pp 1–12. https://doi.org/10.1145/2556464.2556467

A Multi-Biometrics System Based on the IRIS Using Convolutional Neural Networks

Buddesab, K. C. Rajani, G. Geetha, and Syed Hayath

Abstract The iris biometric is about the examination of examples on this surface. It is imperative to guarantee security biometrically in the most verification—and distinguishing-proof situation. The novel examples are utilized in iris acknowledgement to recognize people who require an elevated degree of safety. It investigates a productive strategy that utilizes convolutional neural networks (CNN) to include extraction and arrangement individually to expand the effectiveness of acknowledgment. The action is conceivable in a couple of stages. Picture obtaining, preprocessing, division, standardization, extraction, and matching are various phases of iris acknowledgment. The proposed procedure has been effectively applied, and furthermore, it clearly shows the presentation of trial assessment on iris pictures from the MMU iris data set.

Keywords Multimodal biometric frameworks · Deep learning · Convolutional neural network · MMU iris dataset Gabor channel technique

---

Buddesab (✉) · K. C. Rajani
Department of Engineering, APJ Abdul Kalam School of Engineering,, Garden City University, Banaglore, India
e-mail: buddesab@gcu.edu.in

K. C. Rajani
e-mail: Rajani.kc@gcu.edu.in

G. Geetha · S. Hayath
Department of Artificial Intelligence and Machine Learning, Cambridge Institute of Technology, Bengaluru, KAR 560036, India
e-mail: geetha.aiml@cambridge.edu.in

S. Hayath
e-mail: syedhayath.aiml@cambridge.edu.in

A. Kumar et al. (eds.), *Trends in Mechatronics Systems*, Emerging Trends in Mechatronics, https://doi.org/10.1007/978-981-97-9108-8_10

1 Introduction

In the biometric market, iris biometric recognition systems are the most efficient and widely used. Because the pattern variability of each individual is enormous, iris patterns have been discovered to be an alternative solution for reliable visual recognition of a person. Because of its non-uniform texture, iris features are more distinguishable when compared to other biometric modalities. At the same time, the iris recognition process has been discovered to be more complex. Because of this, the iris recognition process is broken down into steps such as iris image acquisition, preprocessing, feature extraction, and classification of Yuan Zhuang [1]. According to the studies, more than one billion people worldwide have already enrolled their iris images in various databases.

The main known challenges in iris segmentation are due to these databases. As a result, the iris trait is more likely to take the lead in the next generation. The iris recognition system's success is based on the evolution of a dynamic feature descriptor, primarily in the Gabor phase quadrant, which was a remarkable innovation in iris recognition. Better performance can be obtained in comparison to this Gabor phase quadrant approach by developing a new feature representation procedure that improves iris recognition accuracy while reducing computational complexity. One such revolution is the application of the most recent deep learning advances to identify a better feature representation method. Deep learning techniques are demonstrated using a convolutional neural network (CNN) of Wu Haibing [2].

Iris recognition collects about 240 biometric features, each of which is unique to each eye. The system then creates a virtual representation of the iris using the collected data.

Using the most appropriate localization approach, the input eye picture may be reliably recovered from the iris area, then normalized and enhanced using different preprocessing.
It can do classification from the localized iris pictures, and the convolutional neural network can learn unique
The proposed system's performance has been tested on a variety of datasets, including MMU.

2 Literature Survey

The iris identification technology presented by Arora et al. [3] validates an individual's identity using a deep learning approach. The method used the Circular Hough trans-form to localize the iris. A Convolutional Neural Network (CNN) was used to automatically extract features from the localized iris input pictures. A Soft-Max classifier was then applied to classify the dataset into one of 224 classes. Researchers found that the selection of hyper-parameters and optimization algorithms has a major role in the accuracy of their suggested system. They were able to get a 98% accuracy rate by fine-tuning these components of Buddesab [4]

Menon et al. [5] discuss of benefits of utilizing issues When contrasted with hand-created highlights in powerfully delivered elements to take care of both ID and confirmation view of tweaked profound remaining organization models. They adjusted a biometric validation network intended for an alternate occupation to the issue. At long last, as well as improving exactness, CNN might be utilized to really eliminate specific shortcomings in existing biometric frameworks. More concentration on the effective utilization of profound organizations in customary biometric applications, like iris acknowledgment frameworks, is required for outcomes concerning accuracy and security of Buddesab [6].

Nguyen et al. [7] explored the trial of iris authentication, and their studies have revealed that CNN features originally designed for object authentication can indeed be effectively utilized for the iris authentication challenge. They utilized different pre-trained models like AlexNet, VGGNet, InceptionNet, ResNet, ThickNet, and Daugman, and the most vital top affirmation fineness is satisfied by DenseNet among every one of the five CNNs. In two gigantic iris, 2013 and CASIA-Iris-thousand, they satisfied affirmation flawlessness.

The researchers fostered a start-to-finish profound learning design for iris identification by tweaking a pre-prepared convolutional model utilizing Image Net. This might allow for proficiency with the element's portrayal and direct acknowledgment simultaneously, which is extremely helpful for other biometric acknowledgment circumstances where a couple labeled photographs are accommodated for each class. The proposed approach was applied to the IIT-Delhi Iris dataset, yielding promising results that outperformed alternative methods for handling these datasets. Additionally, they demonstrate the use of visualization to identify the most significant iris regions, which greatly influence the recognition outcome of Minaee [8].

Umer et al. [9] proposed Iris distinguishing proof framework in light of a multi-scale morphological element extraction strategy the iris qualities are addressed as the amount of divergence buildup utilizing the morphologic formal hat change. Multi-class issues are reduced to a two-class issue for grouping utilizing the division approach. The proposed framework's presentation is assessed using four iris information bases: IITD, MMU1, UPOL, and UBIRIS [15].

An original iris coding approach in light of varieties in the covering precise fix coefficients of discrete cosine change (DCT) from standard pictures has been proposed. Munro applies his proposed model to the 2156 photographs in the CASIA dataset and the 2955 pictures in the shower data set, yielding a dazzling 100% correct recognition rate (CRR). Besides, a new most pessimistic scenario metric for genuine gadget productivity is given without a trace of matching disappointments and a low Equal Error Rate (EER) of Monro [10, 16]

3 Proposed System

To begin, a preprocessing approach can be utilized to precisely distinguish the iris region from the foundation and all unessential qualities like the understudy, sclera, eyelids, eyelashes, and specular reflections, utilizing an effective and computerized iris confinement.

The essential inspiration for involving the iris district as a contribution to CNN as opposed to the whole eye picture in this study is to bring down the computational intricacy of CNN & Another explanation is to prevent the presence of eyelids and eyelashes from corrupting the coordination and include extraction activities. Following distinguishing proof, the iris region is transformed into a standardized structure with set aspects, permitting direct examination of two iris pictures of different sizes. By applying the CNN as a mechanized include extractor, the standardized iris picture is additionally utilized to serious areas of strength to make interesting iris attributes.

Finally, the matching scores of the right and left iris pictures are joined to conclude the conspicuous confirmation of the person whose iris pictures are being dissected. Different CNN courses of action are ready on the readiness set and surveyed on the endorsement set during the arrangement stage to find the best one with the base mix-up. In the testing stage, its display on test data is surveyed.

4 Implementations

4.1 Biometric Systems and the Processes Involved

The training and testing stages are the two major steps of a biometric system. During the enrollment step, pictures are gathered from a single biometric system and processed to create a clear image, remove distortions, and retrieve the region of interest for extracting features. The pre-processed image's details are extracted through the feature extraction module.

The feature vector is formed by removing certain characteristics from a picture function stored in the database. Figure 1 illustrates the condition. Enrollment mode is a stage in the learning process when you collect biometric data about the people you want to identify. To increase the picture quality during the recognition phase, the query image is pre-processed before testing.

The extracted features are then compared against the feature database within a matching module to generate a matching score. The decision module for a match score was used for detecting authorized individuals. Figure 2 shows these step sequences. There are two functions provided by a biometric system: verification and identification. A 1:1 (one-to-one) search is for verification, whereas a 1:N (one-to-many)

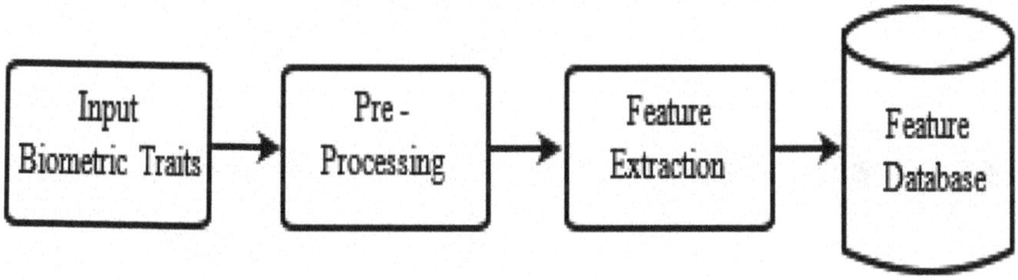

Fig. 1 Steps of the biometric system

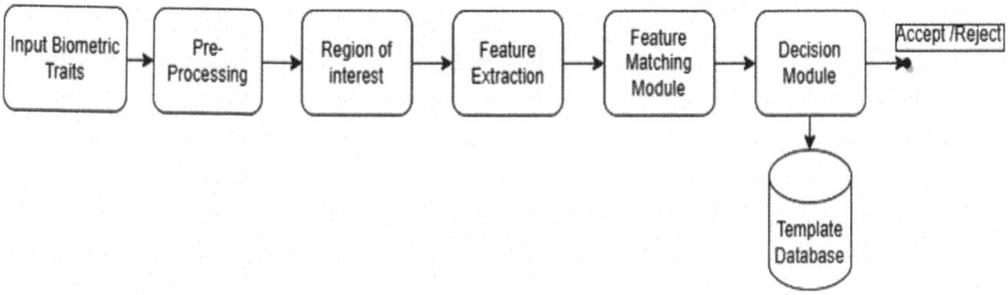

Fig. 2 Steps of the biometric systems for recognition

search is for identification. The biometric system works by claiming the correct verification. In the case of identification, a person tries to prove that the person is the one listed in the user database.

4.2 Processes Involved in IRIS System

Iris images were preprocessed in three phases: enhancement of the image, segmentation of the iris, and normalization of the iris area. However, before segmenting the input photos, the collected images were treated to remove noise during the capture process. The present research also involves changing the image colors, making it a gray-scale image using a median filter, to improve the image's quality as its brightness is not spread equally and to eliminate the noise near the pupil area, as depicted in Fig. 3.

4.3 Segmentation

Identifying the iris section of the eye from the other portions of the retina and determining the iris' outer and inner limits is crucial. However, the procedure becomes

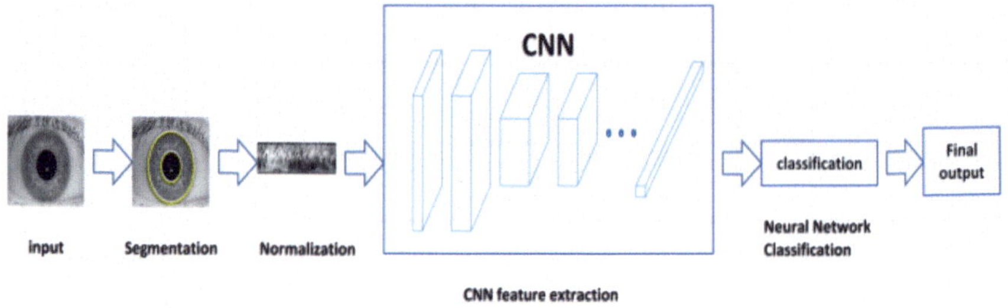

Fig. 3 Processes involved in IRIS system

more difficult when eyelids and eyelashes hide the iris area. Changes in illumination conditions may also impair the quality of the region of the extracted iris during acquisition, affecting the iris' segmentation and the identification results [11].

According to the researchers, both the outer and inner margins of an iris area should be detected with an automated technique. According to the researchers, the outer and inner margins of an iris should be detected with an automated technique. To improve the contrast between the sclera and the iris area, both outer and inner edges were detected by applying an effective technique of improvement based on a multiple Gaussian filter with a histogram process [12, 13]. Along with this, a robust MMU iris dataset is used for determining the pupil's radius and center, along with the iris circle.

4.4 Normalization

Since they were taken from various people and situations, eye images may be of varying sizes. Data is normalized from the same or different sizes to obtain more accurate iris recognition. This is performed by converting the iris area into a rectangular shape. In a segment iris area, every pixel in (x, y) Cartesian coordinates is changed into (r, θ) polar coordinates, with r interval [0,1] and on angle $[0, 2\pi]$ via the rubber sheet algorithm.

4.5 Feature Extraction Using Deep Leaning

Following the creation of iris extraction aspects, normalizing with classification was applied with a deep learning model that combines CNN models with learning algorithms. Figure 4 illustrates the proposed CNN structure. There are seven levels to it. Because CNN works with an input image of fixed size, every training sample was resized to 64360 pixels.

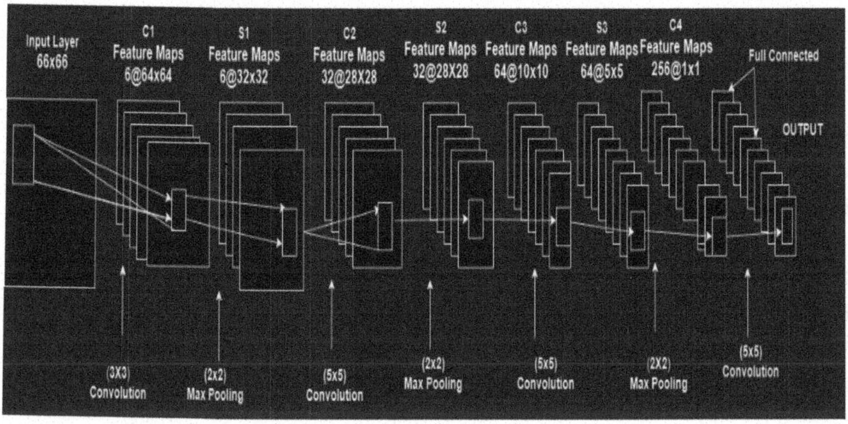

Fig. 4 Model of the iris recognition

CNN layers are divided into two categories: the first four layers are used for extracting features, while the next three hidden levels are responsible for feature evaluation.

In the convolution Layer, the L1: The first layer has a set of 3 × 3 filters. By filter convolution, the input provided 6@64 × 360, and this layer implements the convolutional process. The features map at the output is created using a workable bias and nine workable weights, resulting in a layer with 60 trainable parameters ((6 × 9) + 6). Including Kernel Size of 22 in x and y distance provides 6-feature maps, each of 179 × 32 size, with Max-pooling based on overlapping. The convolutional layer is also Layer 2 (L2). The layer comprises 1760 trainable parameters, with an input picture of 32 kernels: 32@30 × 177 and 9 workable weights with workable bias as an output feature map. The maximum pooling, based on non-overlapping and includes 2 × 2 kernel sizes in x and y distances, yields 32 images feature of size 15 × 88 each.

Layer 3 (L3) comprises 2 layers, similar to Layer 2. The convolutional layer uses a 3 × 3 bit to create (64) features of every dimension 13 × 86. In the max-pooling layer, this is reduced to 6 × 43 pixels, and 64 feature maps are added to L3. L4: (Layer 4) comprises L2 & L3. For every 4 × 41, all 256 features are created using 3 × 3 Bits in a convolution layer.

The identification of characteristics requires the next three main levels. A total of 1024 neurons were used in the first hidden layer. L2, with 512 neurons, has been set up. The third hidden layer has 80 neurons that represent the total classes in input data. An ADAM optimizer working on 32 batch sizes, 0.0005 decay, and 0.0001 learning rates has been implemented into the system. The loss function is frequently measured using methodological cross-entropy, and network variables were changed to lessen the prediction loss. Table 1 shows complete details of the CNN layers of Daugman [14].

Table 1 The complete definition of CNN layers

Layer (type)	Output shape	Param #
conv2d (Con2D)	(None,256,256,64)	4864
max_Pooling2d (MaxPooling2D)	(None,128,128,64)	0
conv2d_1 (Conv2D)	(None,128,128,32)	51,232
max_Pooling2d_1 (MaxPooling2D)	(None,64,64,32)	0
conv2d_2 (Con2D)	(None,60,60,16)	12,816
Dropout (Dropout)	(None,60,60,16)	0
Flatten (Flatten)	(None,57,600)	0
Dense (Dense)	(None,1024)	58,983,424
Dropout_1 (Dropout)	(None,1024)	0
Dense_1 (Dense)	(None,20)	20,500

4.5.1 Convolutional Layer

The parameters (weights) of this layer are made up of collections of learnable kernels generated randomly, and training was done with the back-propagation technique. The used kernels contain some local connections but generally interact through the preceding layer's whole depth.

4.5.2 Max-Pooling Layer

This layer decreases the spatial size of the output from the convolution layer and creates limited translational invariance. Only the approximate position of a given feature relative to other features is preserved once convolution layers detect it Wu Haibing [2].

4.5.3 Fully Connected Layers

The max-pooling output or last convolutional layer is input into fully connected layers similar to the working of a traditional neural network.

5 The Rectified Linear Unit (ReLU) Layer

After every FC layer and Convolutional layer, the ReLU layer is used as an activation function (excluding fully connected layer) in the entire CNN model.

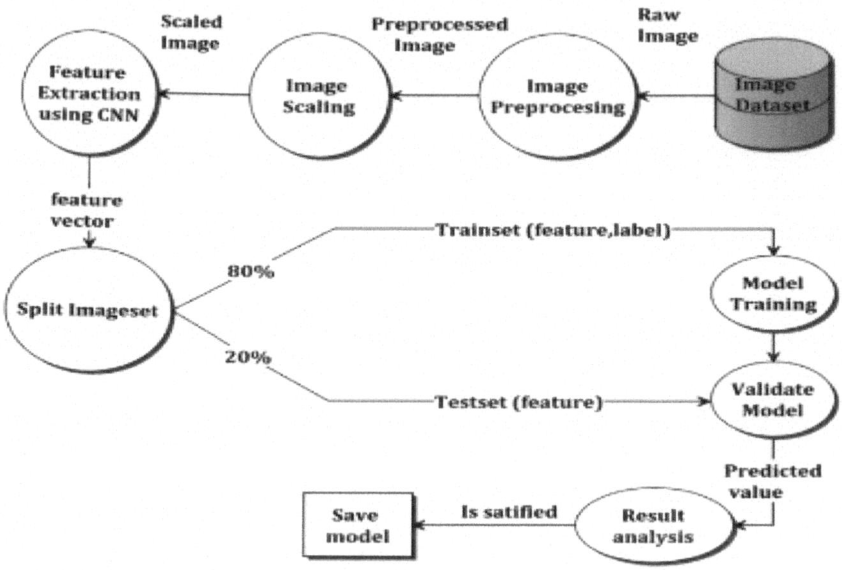

Fig. 5 Methodology

5.1 Preparing Methodology

Through this review, a large portion of the trials were performed using (80%) physically chosen tests for preparation, while the excess (20%) for testing gave a particular arrangement of test information. The fundamental stages of the proposed technique for preparing are characterized in calculation 1, as shown in Fig. 5.

Algorithm 1: CNN Algorithm "//**Input**: The input (normalization images)
//**Output**: The recognition accuracy

1. Separate the photos into two groups: training and validation.
2. Train the CNN configuration using the training set.
3. Validate CNN configuration with a series of validations".
4. Repeat steps 3–4 with *N* epochs in between.
5. Use the first hidden layers to extract features.
6. Use the training set to obtain training labels.
7. Training a multi-class (CNN layers) with the training features.
8. Use a test set to extract features.
9. Evaluate each label for the check set using the trained classifier.
10. Assign known labels to the check set.
11. Represents the expected accuracy.

```
]:  1 test_acc=img1_pred[0][img1_class]

]:  1 print('Accuracy of testdata set : ', test_acc)

    Accuracy of testdata set :  0.9999391
```

Fig. 6 Accuracy for test data

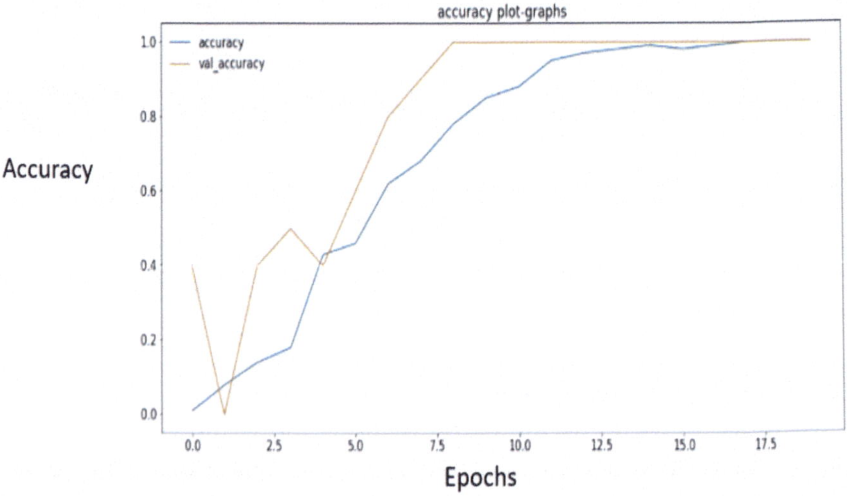

Fig. 7 Accuracy plot graph

6 Results and Analysis

Since the training, the developed system has achieved 0.99% accuracy for both the right and left MMU-Iris datasets. The accuracy as shown in Fig. 6

Both iris pictures are captured in these files under a variety of conditions, including pupil enlargement, closure of the eyelids and eyelashes, and a little shadow. The result shows the recognition accuracy of an elected database of iris images. The graphs shown Figs. 7 and 8, based on the epochs plotted, the accuracy and loss graphs.

7 Conclusions

The development of an advanced multi-modal biometric system for iris identification, combining both the right and left iris of an individual, demonstrates a significant advancement in biometric verification technology. The integration of several

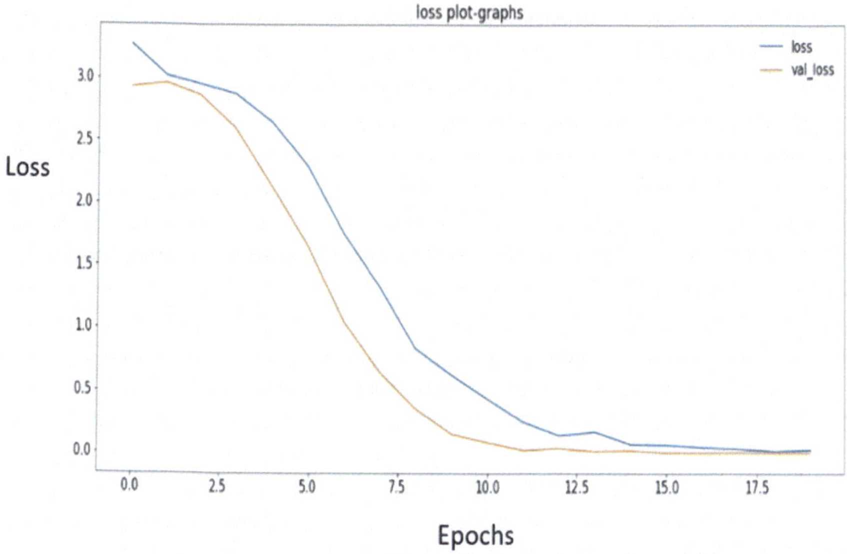

Fig. 8 Loss plot graph

critical stages—segmentation, normalization, feature extraction, and classification—ensures a robust and reliable identification process. The use of Convolutional Neural Networks (CNNs) with transfer learning is a pivotal component in achieving high accuracy. By leveraging pre-trained models, the system benefits from already learned features, thereby enhancing the efficiency and effectiveness of the feature extraction process. This approach, combined with the multi-class CNN layers for classification, results in an impressive accuracy rate of 99%.

References

1. Hitchcock FL (1941) The distribution of a product from several sources to numerous localities. J Math Phys 20:224–230
2. Zhuang Y, Chuah JH, Chow CO, Lim MG (2020) Iris recognition using convolutional neural network. IEEE 10th International conference on system engineering and technology (ICSET), 9 November, Shah Alam, Malaysia
3. Wu H, Gu X (2015) Max-pooling dropout for regularization of convolutional neural networks. In: Neural information processing: 22nd International conference, ICONIP 2015, Istanbul, Turkey, November 9–12, Proceedings, Part I 22, Springer International Publishing. pp 46–54
4. Arora A, Chakraborty P, Bhatia MPS (2020) Analysis of data from wearable sensors for sleep quality estimation and prediction using deep learning. Arab J Sci Eng 45:10793–10812
5. Buddesab T, Thriveni J, Venugopal KR (2019) Trust model genetic node recovery based on cloud theory for underwater acoustic sensor network. Int J Elect Comput Eng 95:3759–3771
6. Menon H, Mukherjee A (2018) Iris biometrics using deep convolutional networks. In: 2018 IEEE International instrumentation and measurement technology conference (I2MTC), IEEE pp 1–5

7. Buddesab et al. (2019) Integration of wireless sensor network and cloud computing using trust and reputation technique. Emerging research in electronics, computer science and technology: proceedings of international conference, ICERECT 2018. Springer Singapore
8. Nguyen K, Fookes C, Ross A, Sridharan S (2017) Iris recognition with off-the-shelf CNN features: a deep learning perspective. IEEE Access 6:18848–18855
9. Minaee S, Minaei M, Abdolrashidi A (2021) Deep-emotion: facial expression recognition using attentional convolutional network. Sensors 21(9):3046
10. Umer S, Dhara BC, Chanda B (2015) Iris recognition using multiscale morphologic features. Pattern Recogn Lett 65:67–74
11. Monro DM, Rakshit S, Zhang D (2007) DCT-based iris recognition. IEEE Trans Pattern Anal Mach Intell 29(4):586–595
12. Daugman J (2007) New methods in iris recognition. IEEE Trans Syst Man Cybern Part B. https://doi.org/10.1109/TSMCB.2007.903540
13. Ma L, Wang Y, Tan T (2002) Iris recognition using circular symmetric filters. In: 2002 International conference on pattern recognition, IEEE vol 2, pp 414–417
14. Daugman J (2009) How iris recognition works. In: The essential guide to image processing, Academic Press. pp 715–739
15. Kumar A, Gulati V, Kumar P (2018) Investigation of process variables on forming forces in incremental sheet forming. Int J Eng Technol 10(3):680–684
16. Sehrawat S, Kumar A, Prabhakar M, Nindra J (2022) The expanding domains of 3D printing pertaining to the speciality of orthodontics. Mater Today Proceed 50:1611–1618

Multi-objective Moth Flame Optimization for Renewable Energy Integration

Sundaram B. Pandya, Kanak Kalita, Ajay Kumar, Pradeep Jangir, and Laith Abualigah

Abstract Investigating the optimization of power flow within an energy grid enriched with renewable sources wind, solar, and hybrid solar-hydroelectric systems this study conducts a techno-economic assessment to determine the most efficient energy distribution under varied scenarios. The challenge of calculating optimal power flow (OPF) is heightened by the unpredictable output from renewable resources. Our approach leverages Lognormal, Weibull and Gumbel distributions

S. B. Pandya
Department of Electrical Engineering, Shri K.J. Polytechnic, Bharuch 392001, India

K. Kalita
Department of Mechanical Engineering, Vel Tech Rangarajan Dr. Sagunthala R&D Institute of Science and Technology, Avadi 600062, India

University Centre for Research and Development, Chandigarh University, Mohali 140413, India

K. Kalita
e-mail: drkanakkalita@veltech.edu.in

A. Kumar (✉)
Department of Mechanical Engineering, School of Engineering and Technology, JECRC University, Jaipur, Rajasthan, India
e-mail: ajay.kumar30886@gmail.com

P. Jangir
Department of Biosciences, Saveetha School of Engineering, Saveetha Institute of Medical and Technical Sciences, Chennai 602105, India

Jadara Research Center, Jadara University, Irbid 21110, Jordan

L. Abualigah
Computer Science Department, Al al-Bayt University, Mafraq 25113, Jordan

MEU Research Unit, Middle East University, Amman 11831, Jordan

Applied Science Research Center, Applied Science Private University, Amman 11931, Jordan

Centre for Research Impact and Outcome, Chitkara University Institute of Engineering and Technology, Chitkara University, Rajpura, Punjab 140401, India

Artificial Intelligence and Sensing Technologies (AIST) Research Center, University of Tabuk, Tabuk 71491, Saudi Arabia

A. Kumar et al. (eds.), *Trends in Mechatronics Systems*, Emerging Trends in Mechatronics, https://doi.org/10.1007/978-981-97-9108-8_11

to forecast power generation, providing a quantitative basis for managing uncertainty. The objective functions that penalize underestimations and allocate reserves for overestimations, fostering financial prudence and encouraging accurate modeling. Utilizing a non-dominated Multi-objective Moth Flame Optimizer integrated with a fuzzy decision-making framework, optimization seek to balance supply and demand efficiently. The effectiveness of this method is validated against three recent algorithms applied to the hybrid Renewable Energy Resources (RER) using the IEEE-30 bus system as a benchmark. Result findings offer valuable insights into advanced strategies for managing renewable energy flows, highlighting potential improvements in grid reliability and economic efficiency.

Keywords Optimal power flow · Renewable energy · Probabilistic forecasting · Moth flame optimization · Techno-economic analysis · Grid integration · Multi-objective optimization

1 Introduction

Electrical networks are the backbone of modern society, powering industries, homes and infrastructure while ensuring economic development and quality of life. The rise of optimal power flow (OPF) principles has been instrumental in improving the performance, reliability, and safety of these networks. OPF methods were introduced more than half a century ago following which they have changed their form and substance to fit changing face of energy world; with this change came a change in control strategies for electrical networks. This change has manifested itself in the way electrical networks are managed and operated, with OPF as a key driver toward improved performance within defined limits aimed at maintaining sustainability in these complex systems. Over the years there has been a steady shift to renewable energy sources from traditional fossil fuel-based methods in the energy sector [1]. Among other challenges due to the intermittent nature of renewable resources such as winds or solar power when being ushered into such systems that sometimes requires innovative OPF solutions that integrate them well into the grid smoothly maintaining stability and efficiency at all times. This field has been active scientifically discussing about connecting renewable energies integration with OPF analysis. Lately several stochastic as well metaheuristic approaches for handling this aspect have emerged showing growing need for sustainable efficient strategies in this direction [2]. This literature review presents diverse methodologies employed so far including new research on optimal power flow reflecting dynamicity characterizing advancements made over time.

Management and operation of an electric network [sic] hinges on optimal power flow (OPF), one basic feature that determines the overall performance and security of the network. OPF is aimed at maintaining a stable and secure operation of an electric network maximizing objective(s) under constraints. The research topic has been a subject of thorough inquiry since its first publication by Carpentier [1] in the last fifty years.

In today's energy landscape, renewable energy sources are becoming increasingly major components of electric systems. However, due to the uncertain nature of these non-conventional energy sources, it is important to carry out rigorous OPF analysis as they introduce a new layer of complexity in an already complex system. Researchers around the world are studying how renewable energy can fuse into OPF with hopes that it will be both effective and sustainable. In this regard, different stochastic approaches have been proposed to solve the OPF problem over time such as those by. It should be noted however that while some metaheuristics might give impressive results for one set of problems; they will not do well on other occasions as claimed by [2].

During the last decade, a plethora of methodologies were developed to solve the OPF problem each addressing a different facet. Ming et al. [3] provided a detailed survey on metaheuristic techniques which were used in solving complexities associated with OPF. Their work also highlights how this field is moving and challenges faced in developing ever more effective and efficient methods for managing evolving power distribution networks.

Most studies cited so far mainly focused on conventional power plants; however, there have been few recent researches where electrical systems containing both wind generators and coal-fired units were analyzed in order to obtain minimum cost of generation. For example, Roy and Jadhav [4] adopted an optimization technique called Gbest guided Artificial Bee Colony (GABC) algorithm in order to improve on their earlier results obtained from similar experiments but with different constraints [4]. These researchers suggested an innovative design for DFIG, which gives specific restriction regarding reactive power (VAR) output capabilities [5, 6]. They further supplemented this with static synchronous compensator (STATCOM), which was considered as a VAR power compensating equipment located within the domain comprising coal-based and wind units. The two approaches used in solving the OPF problem include ACO and the MBFA algorithm. Another important study in this area was proposed by Shi et al. where a new framework to estimate wind turbines generation cost was given. A key challenge they identified for utilities with wind and coal-based units is the need to strategically schedule generators for economic dispatch. Jabr and Pal [7] provided a stochastic model of wind power generation to address this issue. A 2011 study by Mishra et al. [8] employed the DFIG wind turbine model to deal with a similar challenge. Zhou et al. [9] proposed a Dynamic Economic Dispatch (DED) paradigm that considered various constraints related to wind energy and reserve risks. Simultaneously, Dubey et al. [10] integrated the impact of valve-point loading and emissions from the generating unit into the DED structure. A detailed OPF scheduling approach for a unique hybrid network comprising a solar photovoltaic (SPV), battery and diesel generator is delineated by Tazvinga et al. [11]. Kusakana [12] proposed an alternative storage solution for an autonomous hybrid network comprising a wind turbine, SPV and diesel generator, featuring pumped hydro storage as a key element. Remarkably, the problem architecture involves three different types of sources and discusses a single-objective dynamic Economic

Dispatch, as referenced in [13]. Salkuti et al. Day-ahead thermal and renewable power generation scheduling considering uncertainty, [14, 15] contributed to optimal scheduling using unconventional energy sources such as coal-based, solar-wind power and battery storage.

A contemporary electrical energy system that capitalizes on renewable energy sources introduces a multi-objective OPF (OPF) problem, solvable using the NSGA-II optimization method, providing alternative solutions based on simulation outcomes. Pandya and Jariwala [16, 17] analyze newly developed metaheuristic algorithms for both single and multi-objective OPF issues integrating various renewable energy sources. Biswas et al. [18] presented a system that utilizes varying probability density functions (PDFs) for solar, wind and run-of-river units to evaluate the OPF problem, signifying a continuing evolution of strategies to optimally manage the flow of power from an increasingly diverse array of energy sources.

In the evolving landscape of energy generation and distribution, there remains a considerable need for in-depth research into both conventional power plants and renewable energy systems. The current investigation highlights the incorporation of coal-based, solar, wind, and small hydro units in both single and multi-objective optimum power flow (MOOPF) contexts. The variability of wind and solar power probability density function (PDF) characteristics, coupled with varying levels of penetration, result in a wide range of optimal pricing scenarios.

The highlights of this research are—

1. This study focuses on the numerical modeling of both OPF issues, incorporating a blend of conventional and renewable energy units.
2. It offers a comprehensive analysis, considering four different scenarios that combine renewable energy sources with traditional coal-based power plants.
3. The study employs suitable PDFs to model the stochastic behavior of solar photovoltaic (SPV), wind and small hydro power plants, capturing their inherent unpredictability.
4. The MOMFO technique is leveraged to solve both single and multi-objective OPF problems, particularly in the context of stochastic RES's.
5. Results are meticulously compared across multiple iterations of three newly developed methods, providing valuable insights into the efficiency of the proposed strategies.

The rest of the article is structured as follows: Sect. 2 explores mathematical models, providing an in-depth look at the representation of uncertainty in solar, wind, and small hydro energy outputs within the context of the OPF problem. Section 3 explores the objectives of optimization, while Sect. 4 elaborates on the application of the Multi-objective Moth Flame Optimization technique. Section 5 presents the numerical results and subsequent discussion. Finally, Sect. 6 draws conclusive remarks and insights derived from the research.

2 Mathematical Models

Data associated with this study, the different scenarios for the modified IEEE-30 bus, which incorporates both traditional coal-based power plants and sustainable energy sources.

2.1 *Price of Coal-Based Power Units*

The output power of coal-based power plants, which rely on fossil fuels, can be modeled using the quadratic curve provided below.

$$C_{TO}(P_{TG}) = \sum_{i=1}^{N_{TG}} a_i + b_i P_{TGi} + c_i P_{TGi}^2 \tag{1}$$

To make this model more applicable to real-world scenarios, the effect of the valve point is also included.

$$C_T(P_{TG}) = \sum_{i=1}^{N_{TG}} a_i + b_i P_{TGi} + c_i P_{TGi}^2 + \left| d_i \times \sin\left(e_i \times \left(P_{TGi}^{\min} - P_{TGi}\right)\right)\right| \tag{2}$$

The relevant values for price and emission coefficients associated with coal-based energy generation are displayed across various scenarios. This detailed representation allows for a thorough understanding of the interplay between different energy sources, cost implications and environmental impacts, thereby providing a comprehensive perspective on OPF scenarios.

2.2 *Direct Price of Stochastic Sustainable Plants*

Due to their reliance on environmental factors, renewable energy sources are naturally subject to a certain degree of unpredictability. Thus, cost analysis for these sources includes three main components: direct costs, reserve costs and penalty costs, as outlined by Biswas et al. [18]. For wind farms, the direct costs are modeled using the P_{ws}, the scheduled power from the wind sources.

$$C_w(P_{ws}) = g_w P_{ws} \tag{3}$$

In this equation, g_w represents the direct price coefficient and P_{ws} denotes the scheduled power of the wind energy unit. Similarly, for solar photovoltaic (SPV) farms, the direct costs are determined by the P_{ss}, the planned power from solar sources.

$$C_s(P_{ss}) = h_s P_{ss}. \quad (4)$$

In this instance, the direct costs are computed using P_{ss}., the scheduled output from a combination of SPV and hydel power plants

$$C_s(P_{ss}) = C_s\left(P_{ssh,s} + P_{ssh,s}\right) = h_s P_{ssh,s} + m_s P_{ssh,h}. \quad (5)$$

In this model, P_{ss} represents the total scheduled output, where $P_{ssh,s}$. is the contribution from the SPV plant and $P_{ssh,s}$ is the contribution from the small hydro unit. It is noteworthy that the direct price coefficients *h_s* and *m_s* remain consistent with their prior usage for the *S* units.

2.3 Reserve and Penalty Prices for Sustainable Wind Power Units

The reserve cost for the wind energy unit is defined according to Biswas et al. [18],

$$C_{Rw}(P_{ws} - P_{\mathrm{wav}}) = K_{Rw}(P_{ws} - P_{\mathrm{wav}}) = K_{Rw} \int_{0}^{P_{ws}} (P_{ws} - p_w) f_w(p_w) dp_w. \quad (6)$$

in which K_{Rw} signifies the reserve price coefficient for the wind unit, while P_{ws} represents the available power from the same. In this formulation, $f_w(p_w)$ symbolizes the PDF for energy production in the specified unit. Further, the penalty cost associated with the wind energy unit can be represented by the following equation.

$$C_{Pw}(P_{\mathrm{wav}} - P_{ws}) = K_{Pw}(P_{\mathrm{wav}} - P_{ws}) = K_{Pw} \int_{P_{ws}}^{P_{wr}} (p_w - P_{ws}) f_w(p_w) dp_w. \quad (7)$$

In this representation, K_{Pw}. stands for the penalty price coefcit of the wind energy unit, while P_{wr} indicates the specified output power from the same unit. These coefficients help to account for the potential fluctuations and unpredictability inherent in wind power generation.

2.4 Reserve and Penalty Prices for Sustainable Solar PV Power Units

The reserve cost for a solar photovoltaic (SPV) unit can be modeled using the following equation:

$$C_{Rs}(P_{ss} - P_{\text{sav}}) = K_{Rs}(P_{ss} - P_{\text{sav}}) = K_{Rs}{*}f_s(P_{\text{sav}} < P_{ss}){*}[P_{ss} - E(P_{\text{sav}} < P_{ss})]. \tag{8}$$

In this expression, K_{Rs} symbolizes the reserve price coefficient for the SPV unit, while P_{sav}. refers to the available power from this unit. The term $f_s(P_{\text{sav}} < P_{ss})$ illustrates the likelihood of a shortfall in solar output power in comparison to the scheduled output power (P_{ss}), $E(P_{\text{sav}} < P_{ss})$ indicates the expected SPV output power that falls below the planned P_{ss}.. Moreover, the penalty cost for underestimating the SPV unit's output can be detailed by another equation:

$$C_{Ps}(P_{\text{sav}} - P_{ss}) = K_{Ps}(P_{\text{sav}} - P_{ss}) = K_{Ps}{*}f_s(P_{\text{sav}} > P_{ss}){*}[E(P_{\text{sav}} > P_{ss}) - P_{ss}]. \tag{9}$$

Here K_{Ps} signifies the penalty price coefficient for the SPV unit, $f_s(P_{\text{sav}} > P_{ss})$. displays the chance of the solar output exceeding the scheduled output power (P_{ss}), $E(P_{\text{sav}} > P_{ss})$ reflects the expectation of the SPV output power surpassing the preplanned P_{ss}. These formulas effectively account for potential over and underestimations of power output from SPV units, which can occur due to the inherently unpredictable nature of solar power generation.

3 Reserve and Penalty Prices of Hybrid SPV and Small-Hydro Power

This research focuses on a unique power generation confiration consisting of a small-scale hydroelectric unit and a solar photovoltaic (SPV) unit. The power output from the small hydro unit, which typically depends on the flow rate of the river, is assumed to adhere to the Gumbel distribution in our study. Considering that power generation from the small hydro component only contributes to 10–20% of the total output from the hybrid system, both reserve and using Eq. (8),enalty costs for overestimated and underestimated predictions of the total hybrid system's power production is incorporated. The reserve cost for an overestimated output from the hybrid system can be expressed using the following equation:

$$\begin{aligned} C_{Rsh}(P_{ssh} - P_{\text{shav}}) &= K_{Rsh}(P_{ssh} - P_{\text{shav}}) \\ &= K_{Rsh} * f_{sh}(P_{\text{shav}} < P_{ssh}) * [P_{ssh} - E(P_{\text{shav}} < P_{ssh})] \end{aligned} \tag{10}$$

In this expression, K_{Rsh} is the reserve price coefficient corresponding to the combined system and P_{shav} stands for the accessible power from this system. The term $f_{sh}(P_{shav} < P_{ssh})$ signifies the likelihood of the hybrid system's output falling short of the scheduled output power P_{ssh} is $E(P_{shav} < P_{ssh})$ is the expected output from the system that is below P_{ssh}. The penalty cost for underestimating the hybrid system's output can be described by the subsequent equation:

$$\begin{aligned} C_{Rsh}(P_{ssh} - P_{\text{shav}}) &= K_{Rsh}(P_{ssh} - P_{\text{shav}}) \\ &= K_{Rsh} * f_{sh}(P_{\text{shav}} < P_{ssh}) * [P_{ssh} - E(P_{\text{shav}} < P_{ssh})] \end{aligned} \tag{11}$$

Here K_{Psh} represents the penalty price coefficient associated with the combined system, $f_{\text{sh}}(P_{\text{shav}} > P_{\text{ssh}})$ indicates the possibility of the hybrid system's power surpassing the scheduled power $(P_{ssh}) \cdot E(P_{\text{shav}} > P_{ssh})$ denotes the expected output from the system that exceeds P_{ssh}. the PDF models of renewable energy sources at various locations. For further details regarding the modeling of uncertainties and power generation from renewable energy resources, refer to [18].

4 Objectives of Optimization

This section incorporates various objectives for the optimization of power flow that involves solar, wind, and small-scale hydroelectric plants.

4.1 *Entire Price Minimization*

The principal aim is to reduce costs as much as possible. The total cost is computed as the sum of the unit price for coal-based energy and the direct, reserve and penalty costs associated with the use of non-traditional resources. To combat climate change, many nations are urging the entire electricity sector to lower carbon emissions. A carbon tax (C_{tax}) is gas emissions to incentivize investments in RES. The price of emissions (expressed in dollars per hour) can be calculated as follows:

$$\text{Emission price, } C_E = C_{\text{tax}} E. \tag{12}$$

The overall price for RES can be represented as follows:
Objective 1: Minimize –

$$\begin{aligned} F1 = C_{\text{Tot}} = C_T(P_{TG}) &+ [C_w(P_{ws}) + C_{Rw}(P_{ws} - P_{\text{wav}}) + C_{Pw}(P_{\text{wav}} - P_{ws})] \\ &+ [C_s(P_{ss}) + C_{Rs}(P_{ss} - P_{sav}) + C_{Ps}(P_{\text{sav}} - P_{ss})] \\ &+ \left[C_{sh}(P_{ssh}) + C_{Rsh}(P_{ssh} - P_{\text{shav}}) + C_{Psh}(P_{\text{shav}} - P_{ssh})\right] \end{aligned} \tag{13}$$

The analysis also considers the cost of emissions

$$\begin{aligned} F6 = C_{\text{Tot}} = C_T(P_{TG}) &+ [C_w(P_{ws}) + C_{Rw}(P_{ws} - P_{\text{wav}}) + C_{Pw}(P_{\text{wav}} - P_{ws})] \\ &+ [C_s(P_{ss}) + C_{Rs}(P_{ss} - P_{\text{sav}}) + C_{Ps}(P_{\text{sav}} - P_{ss})] \\ &+ \left[C_{sh}(P_{ssh}) + C_{Rsh}(P_{ssh} - P_{\text{shav}}) + C_{Psh}(P_{\text{shav}} - P_{ssh})\right] + C_{\text{tax}} E \end{aligned} \tag{14}$$

4.2 Emission Minimization

Utilizing coal-based power plants leads to harmful gas emissions, which can be quantified in tons per hour as:

$$\text{F2} = \text{Emission}, E = \sum_{i=1}^{N_{TG}} \left[\left(\alpha_i + \beta_i P_{TGi} + \gamma_i P_{TGi}^2 \right) \times 0.01 + \omega_i e^{(\mu_i P_{TGi})} \right] \text{in ton/hr} \tag{15}$$

4.3 VD Minimization

Bus voltage plays a crucial role in maintaining safety and authority in the system. Keeping the individual unit PQ bus voltage fluctuations at or below 1.0 is key for a satisfactory voltage profile. The success criteria can be stated as follows:

Objective 2: Minimize –

$$F2 = \sum_{i=1}^{N_{pq}} |v_i - 1.0| \tag{16}$$

where N_{pq} shows the number of load (PQ) buses, v_i shows the *p.u* voltage level of *i* th bus.

4.4 APL Minimization

The real power losses P_{LOSS} (MW) can be computed by:

Objective 3: Minimize—

$$F3 = P_{\text{LOSS}} = \sum_{i=1}^{NB} P_{Gi} - \sum_{i=1}^{NB} P_{Di} \tag{17}$$

where P_{Gi} and P_{Di} represent the output and dispatch at *i* th bus; *NB* shows the number of buses.

4.5 Improvement of Voltage Stability Index Including Sustainable Energy Sources

The L_{max} index is the most important, as it symbolizes the voltage stability margin of each bus under standard operating conditions. A scalar value is assigned to each PQ bus by the L_{max}. For the j th bus, the voltage collapse indicator value can be stated as follows:

$$L_j = \left| 1 - \sum_{i=1}^{N_g} F_{ji} \frac{V_i}{V_j} \right| \forall j = 1, 2, \ldots\ldots, NL \tag{18}$$

$$F_{ji} = -[Y_1]^{-1}[Y_2]. \tag{19}$$

where Y_1 and Y_2 were the sub-matrices of Y_{BUS}. The objective function of voltage stability improvement is written by;

$$F4 = L = \max(L_j) \forall j = 1, 2, \ldots\ldots, NL \tag{20}$$

4.6 Equality Constraints

The power flow equations, through their equality constraints, emphasize a fundamental principle in electrical systems: the real and imaginary power produced within the system must not only account for the immediate consumption demands but also compensate for any energy losses that may occur during transmission or distribution. In essence, for a system to function optimally and maintain its reliability, the total power generated should be congruent to the summation of the system's load demand and any associated power dissipation. This ensures that there's neither an excess nor a deficit of power, enabling a balanced and efficient operation of the power grid.

$$P_{Gi} - P_{Di} - V_i \sum_{j=1}^{NB} V_j \left[G_{ij} \cos(\delta_{ij}) + B_{ij} \sin(\delta_{ij}) \right] = 0 \forall i \in NB \tag{21}$$

$$Q_{Gi} - Q_{Di} - V_i \sum_{j=1}^{NB} V_j \left[G_{ij} \sin(\delta_{ij}) - B_{ij} \cos(\delta_{ij}) \right] = 0 \forall i \in NB \tag{22}$$

4.7 Inequality Constraints

Inequality constraints were set by the operation and safety limits of the lines and PQ buses

$$P_{TGi}^{\min} \leq P_{TGi} \leq P_{TGi}^{\max} \forall i \in N_{TG} \tag{23}$$

$$P_{ws}^{\min} \leq P_{ws} \leq P_{ws}^{\max} \tag{24}$$

$$P_{ss}^{\min} \leq P_{ss} \leq P_{ss}^{\max} \tag{25}$$

$$P_{ssh}^{\min} \leq P_{ssh} \leq P_{ssh}^{\max} \tag{26}$$

$$Q_{TGi}^{\min} \leq Q_{TGi} \leq Q_{TGi}^{\max} \forall i \in N_{TG} \tag{27}$$

$$Q_{ws}^{\min} \leq Q_{ws} \leq Q_{ws}^{\max} \tag{28}$$

$$Q_{ss}^{\min} \leq Q_{ss} \leq Q_{ss}^{\max} \tag{29}$$

$$Q_{ssh}^{\min} \leq Q_{ssh} \leq Q_{ssh}^{\max} \tag{30}$$

$$V_{Gi}^{\min} \leq V_{Gi} \leq V_{Gi}^{\max}, i = 1, \ldots.., NG \tag{31}$$

$$\text{Security bounds}: \; V_{L_p}^{\min} \leq V_{L_p} \leq V_{L_p}^{\max}, p = 1, \ldots.., NL \tag{32}$$

$$S_{l_q} \leq S_{l_q}^{\max}, q = 1, \ldots.., nl \tag{33}$$

The equations governing the real power output constraints for various energy-generating units are detailed as follows: Coal-fired units: Eq. (23), Wind-powered units: Eq. (24), Solar-powered units: Eq. (25), and Small hydroelectric units: Eq. (26). Subsequently, the Reactive Power (VAR) capacities of these generating units are defined by a separate set of equations: Coal-fired units: Eq. (27), Wind-powered units: Eq. (28), Solar-powered units: Eq. (29), and Small hydroelectric units: Eq. (30). Within the *NG* system framework, all voltage controlled buses are distinctly depicted. For *PV* buses, their voltage constraints are delineated in Eq. (31). On the other hand, for *PQ* buses, which are typically non-voltage controlled buses, the voltage constraints are given by Eq. (32). Here, "*NL*" denotes the total count of *PQ* buses within the system. Lastly, to ensure the stability and efficiency of power transmission, line loading constraints are crucial. For a system that consists of "*nl*" transmission lines, Eq. (33) outlines the requisite limits that need to be adhered to, guaranteeing the efficient and safe conveyance of electrical power across the network.

5 Multi-objective Moth Flame Optimizer

5.1 Inspiration

This method is fundamentally inspired by the moths undertake lengthy journeys, often spanning great distances, driven by a crucial and sustainable process. This process involves maintaining a specific orientation in relation to the moon, which serves as their reference point. Due to the transverse orientation mechanism's limitations, it facilitates linear travel when navigating using distant light sources. When moths encounter artificial light sources, they strive to maintain a specific angle relative to the source, allowing them to fly in a straight line. This behavior has been the basis for the MFO algorithm developed by Mirjalili [19].

5.2 Multi-objective MFO Algorithm

The algorithm was inspired by the moths' transverse orientation technique, which they use for long-distance navigation in nature. At night, moths can fly straight as they know how to adjust their angle around the moon, a distant light source. However, if the angle is not maintained, they will spiral around the light source and in the case of an artificial light source, they may ultimately be drawn toward it. The MFO method is divided into three stages.

5.3 Formulation of Multi-objective Function

In multi-objective optimization problems, various conflicting objective functions are optimized simultaneously, assuming all constraints are met. Previously, various optimization methodologies were used to solve the OPF problem. Several researchers have addressed multi-objective problems by transforming them into single-objective ones. They achieve this by amalgamating two or more conflicting objectives into a unified objective using a weighted components approach. While this method simplifies the optimization process, it often oversimplifies the complexities inherent to multi-objective problems.

A potentially more nuanced and accurate approach to managing multi-objective challenges is to evaluate the full spectrum of optimal trade-offs. By doing so, one can pinpoint the most favorable compromise solutions within each Pareto front. In multi-objective optimization, the Pareto front represents a set of non-dominated solutions, where no single solution is categorically superior to any other across all objectives. The Nondominated Sorting (NDS) method, when applied to such scenarios, often culminates in two distinct outcomes. The first possibility is that one particular objective asserts dominance, overshadowing the others in terms of its influence on the solution set. Conversely, the second potential outcome is a scenario where no single-objective holds a clear dominance over others. In such cases, the challenge becomes striking a balance among competing objectives, ensuring that the final solutions are representative of an optimal compromise across the entire objective space. This highlights the importance of understanding and navigating the intricacies of multi-objective optimization, rather than relying solely on oversimplified, single-objective transformations. In other words, without compromising specificity, u_1 dominates u_2 only if the following two conditions are met;

$$\forall i \in \{1, 2, 3 \ldots\ldots N\} : f_i(u_1) \leq f_i(u_2) \tag{34}$$

$$\exists j \in \{1, 2, 3 \ldots\ldots N\} : f_j(u_1) \leq f_j(u_2) \tag{35}$$

The application of the Nondominated Sorting (NDS) technique can result in two possibilities: either one objective will dominate over the others, or none of the objectives will dominate. In simpler terms, u_1 dominates u_2 if the following two conditions are met without sacrificing accuracy. If any of the previous conditions aren't met, the configuration u_1 doesn't control u_2 anymore. When u_1 outranks u_2, this is referred to as the non-dominated configuration.

Figure 1 provides a flowchart of the MFO method for solving the OPF problem. Parameters such as population size N_{pop} and stopping value, which represents the maximum number of generations, are introduced alongside the approach. Furthermore, a random parent population P_o is generated within the feasible space, S and each objective function of the objective vector F for P_o is assessed. Following that, as explained in Savsani and Tawhid [20], NDS is applied to P_o, coupled with the computation of crowding distance. The MFO method is then employed to generate a new population P_t, which is subsequently merged with P_o to form the combined population P_i. Through elitism and non-domination, this P_i is organized and the top N_{pop} configurations are updated to create another parent population based on the calculated crowding distance (CD) and non-domination rank (NDR) values. The optimization procedure iteratively progresses, moving through successive generations or cycles. With each iteration, solutions are evolved and refined based on the algorithm's principles. This iterative refinement continues until a predetermined maximum number of generations (or cycles) is achieved. It is crucial to highlight that though in the current study, the termination criteria based on the maximum number of generations is used, there is flexibility in the methodology to use other criteria. Specifically, one can also employ a termination criterion grounded in the cumulative function evaluations, ensuring that the algorithm stops once a particular threshold of evaluations

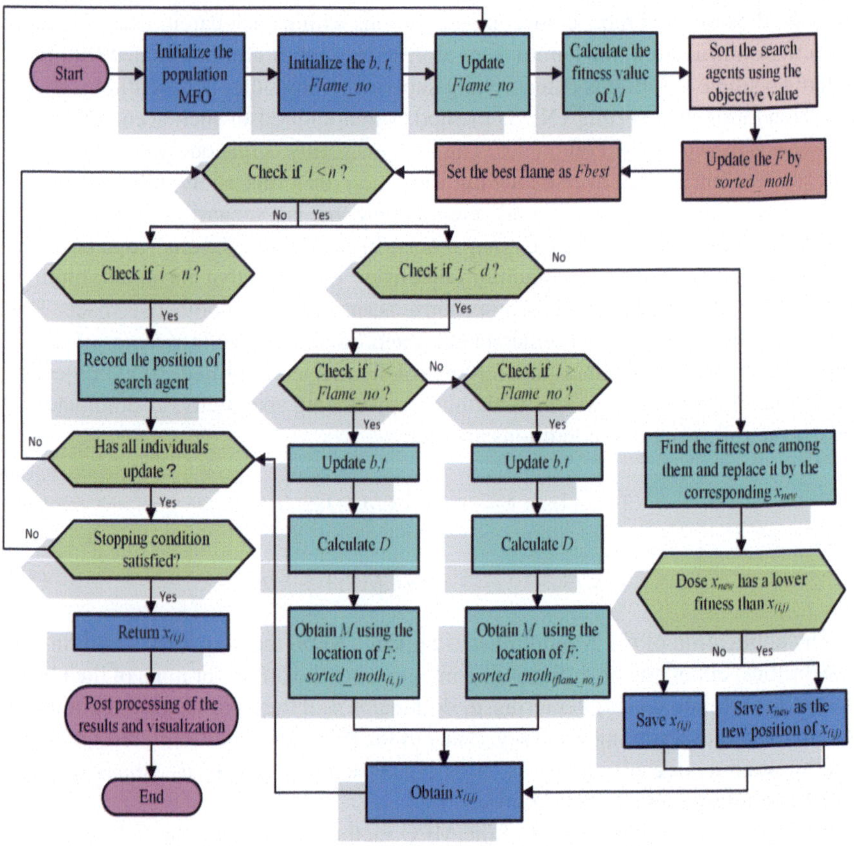

Fig. 1 Diagrammatic representation of projected MFO approach to solve OPF

is surpassed. This alternative end criteria can be particularly valuable in scenarios where the computational expense is a concern, or when the focus is on achieving a predefined level of solution accuracy.

5.4 Best Compromise Solution

In multi-objective functions, the Best Compromise Solution (BCS) can be identified among the non-inferior results by employing the fuzzy membership function technique. The fuzzy membership function μ_{f_i} is expressed as follows:

$$\mu_{f_i} = \begin{cases} 1 & f_i \leq f_i^{\min} \\ \frac{f_i^{\max} - f_i}{f_i^{\max} - f_i^{\min}} & f_i^{\min} < f_i < f_i^{\max} \\ 0 & f_i \geq f_i^{\max} \end{cases} \tag{36}$$

The membership function values range from 0 to 1, signifying the degree of satisfaction of the function f_i The supervisory function μ^k should be calculated as follows. For more in-depth information, refer to Mirjalili [19].

$$\mu^k = \frac{\sum_{i=1}^{N} \mu_{f_i}^k}{\sum_{k=1}^{M} \sum_{i=1}^{N} \mu_{f_i}^k} \tag{37}$$

6 Simulation Results and Analysis

This study encompasses a broad spectrum of eleven diverse test cases. Of these, six exclusively target singular optimization goals. In contrast, the remaining five bring into play a combination of objectives pertaining to both coal and wind energy infrastructures. Computational platform consists of a personal computer fortified with an Intel i5 processor clocked at 3.4 GHz and backed by 8 GB of RAM. The entirety of the programmatic approach was constructed using MATLAB. To ensure robustness, each strategy was subjected to ten separate tests, adjusting the search agent's value to 40 for this specific round of experimentation. MFO method is applied to tackle the stochastic OPF (OPF) problem incorporating three Renewable Energy Resources (RER), namely wind, solar and a combination of solar and small-scale hydroelectric power plants. The MFO technique serves to identify OPF solutions involving sustainable power plants. While resolving the OPF problem, newer methods like Nutcracker Optimization (NO) Abdel-Basset et al. [22], Enhanced Snake Optimizer (ESO) Yao et al. [21], PID-based search Algorithm (PIDBSA) Gao [22], and Osprey Optimization Algorithm (OOA) [23] are also validated for comparison. To showcase the efficacy and viability of proposed methodology, a modified version of the well-established IEEE-30 bus system is employed. This modified plan introduces various forms of renewable energy sources such as wind turbine, solar photovoltaic installations, and mini-hydro power plants. By employing this mix of different energy sources within an IEEE-30 bus system, the objective is to provide a holistic representation that simulates actual utility systems and showcases how it can be used to optimize power distribution and usage. The following four scenarios were considered in this study:

(1) OPF with two wind units and traditional coal-based power plants.
(2) OPF integrating three wind units and traditional coal-based power plants.
(3) OPF combining two wind power units, one solar power unit and traditional coal-based power plants.
(4) OPF with a hybrid power generation system, including wind, solar, small-scale hydroelectric power plants, and traditional coal-based power plants.

6.1 Scenario 1 OPF with Two Wind Units and Coal-Based Power Plants

In the present study, the complex challenges associated with the stochastic optimal power flow (OPF) problem, particularly when integrating renewable wind energy sources alongside traditional coal-based power plants is studied. Utilizing both single-objective and multi-objective optimization frameworks, our primary method of investigation and resolution is the Moth Flame Optimization (MFO) algorithm. To provide a tangible understanding of the proposed methodology, the implement it on a specially adapted IEEE-30 bus system that integrates both wind energy and coal-fueled power units. Critical data points such as line specifics, load profiles, and particulars about the wind energy installations have been sourced from reference [18]. Focusing on the six singular-objective scenarios, the Moth Flame Optimization technique was applied to derive the optimal solutions. The findings of this optimization are systematically collated in Table 1.

For instance, in the context of Case 1 that prioritizes sustainable wind energy solutions, observed a Total Fuel Price (TFP) pegged at 669.6325 $/hr. Figure 2 provides a comprehensive convergence graph depicting the TFP across the four considered methodologies. Transitioning to Case 2, noted an emission rate for pollutant gases standing at 0.092314 Ton/hr. As can move to Case 3, the cumulative power loss experienced across several transmission networks sums up to 1.746933 MW. Case 4 offers insights into the minimal voltage fluctuation at each bus point, measuring at 0.294445 *p.u.* Diving deeper, after calculate the Voltage Stability Index (or the L_max index), which typically oscillates between 0 (indicating no load) and 1 (indicative of peak load, bordering on voltage instability). In the context of Case 5, recorded minimum for the L_max index was 0.134. Finally, for Case 6, with a carbon tax rate (C_tax) fixed at $20 per ton, the comprehensive cost encompassing emission metrics totaled 697.8024 $/hr. When juxtaposed with the results garnered from three contemporary algorithms, it becomes evident that the MFO methodology is particularly adept at addressing the OPF quandary, especially when integrating wind with coal-based energy units. Switching attention to.

Table 1 Comparative analysis of single-objective simulations for scenario 1

Case	Control and state variables	MFO	No	ESO	PIDBSA
Case 1	TFP ($/hr)	669.6325	669.7108	669.721	689.8203
Case 2	Emission (Ton/hr)	0.092314	0.092314	0.092314	0.092314
Case 3	APL (MW)	1.746933	1.750292	1.748425	1.769592
Case 4	VD (p.u)	0.294445	0.298861	0.305163	0.384215
Case 5	L index	0.134	0.134	0.134	0.134
Case 6	TFP With Carbon Tax ($/hr)	697.8024	697.8143	698.1313	701.8083

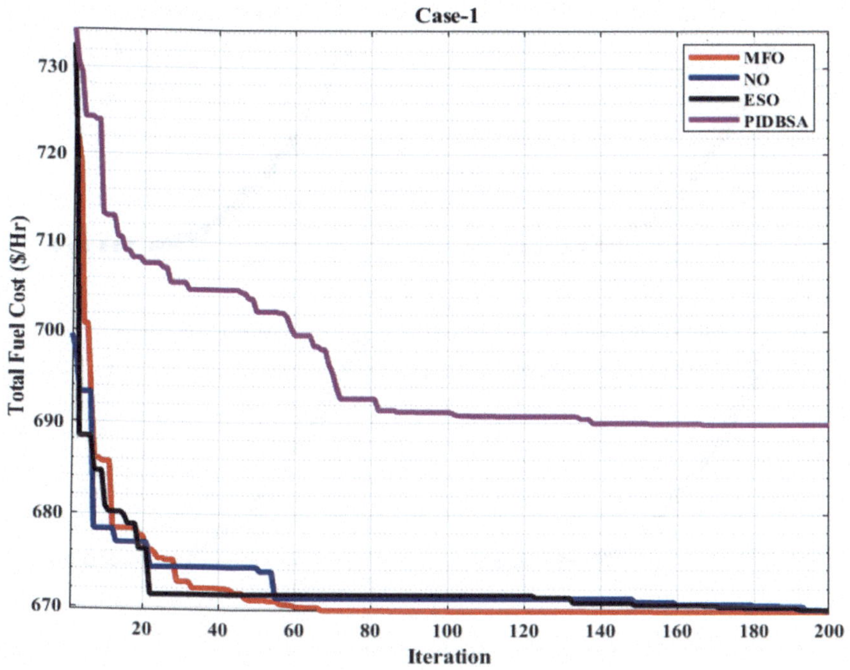

Fig. 2 Convergence of TFP minimization in scenario 1

Cases 7 through 11, these are earmarked as multi-objective optimization challenges. Here, four leading-edge methods, including the MOGNDO algorithm [16], multi-objective sine–cosine algorithm (MOSCA) [24] and the multi-objective slime mold algorithm (MOSMA) [25], are deployed, all striving to simultaneously optimize multiple objectives. Figures 3, 4 and 5 visually encapsulate the best possible trade-off solutions, as discerned using a fuzzy decision-making methodology. The consolidated findings of these multi-objective OPF challenges are showcased in Table 2. An analysis of the outcomes suggests that the Multi-objective MFO (MOMFO) framework is exceptionally proficient at pinpointing the most effective solutions to the multi-objective OPF conundrum that harmoniously integrates wind energy with traditional coal-based power generation units.

6.2 *Scenario 2 OPF with Three Wind Units and Coal-Based Power Plants*

In this advanced scenario [26], the objectives mirror those articulated in Scenario 1. To concurrently optimize them, all four renowned optimization techniques are put to work. The culmination of this rigorous optimization process for individual objectives

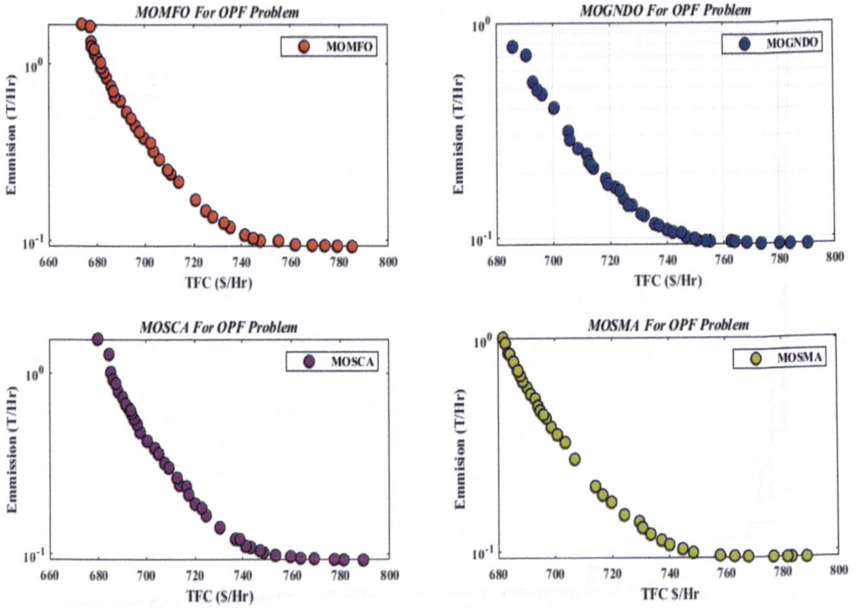

Fig. 3 Pareto front of TFP and emission minimization in scenario 1

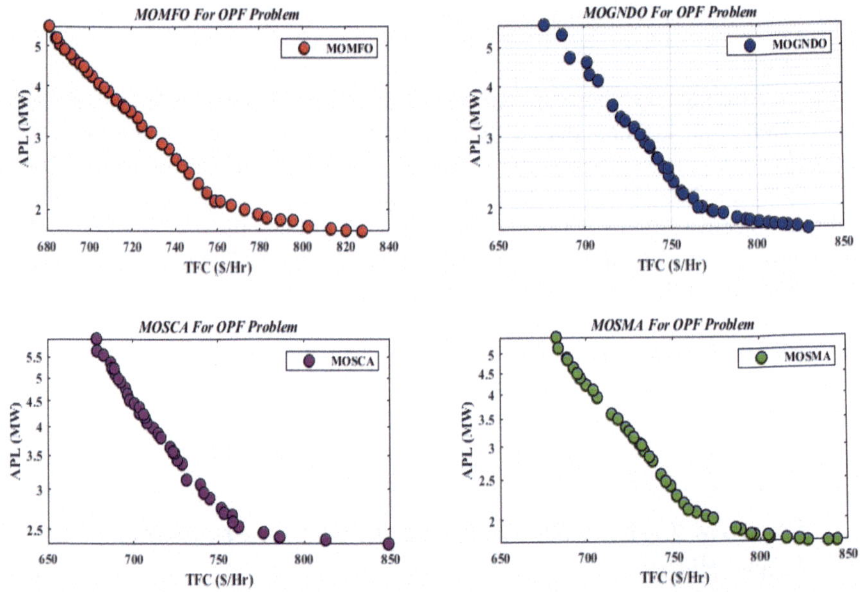

Fig. 4 Pareto front of TFP and APL minimization in scenario 1

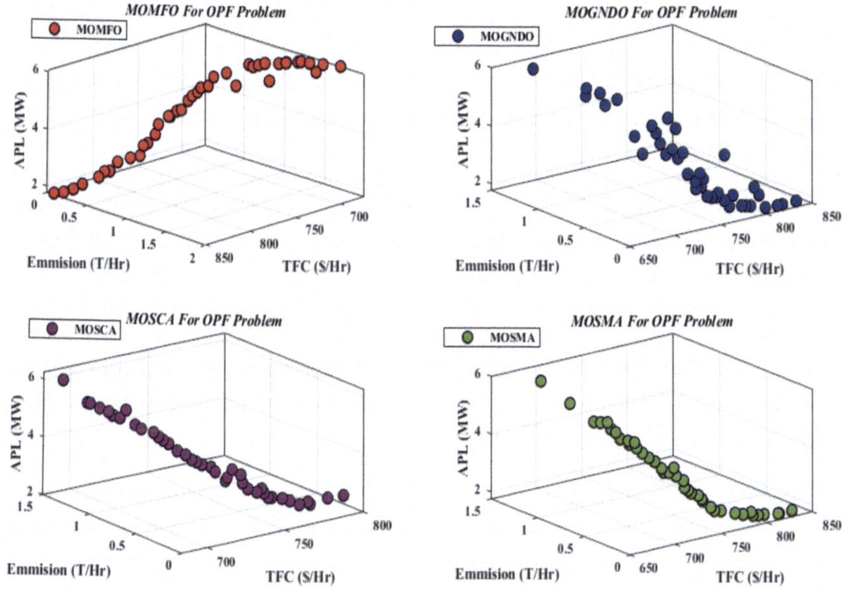

Fig. 5 Pareto front of TFP, emission, and APL minimization in scenario 1

Table 2 Comparative analysis of multi-objective simulations for scenario 1

Problem	Multi-objectives	MOMFO	MOGNDO	MOSCA	MOSMA
CASE 7	TFP ($/hr)	697.7538346	708.7733935	703.4784215	706.9222705
	Emission (Ton/hr)	0.413146137	0.25974756	0.382232276	0.271262352
CASE 8	TFP ($/hr)	728.9709488	728.5983848	728.4949431	732.9203709
	APL (MW)	3.087080567	3.154706801	3.372615386	2.920379886
CASE 9	TFP ($/hr)	737.2466709	735.8515182	722.8742917	730.390622
	Emission (Ton/hr)	0.117380162	0.137769619	0.198433309	0.13782517
	APL (MW)	3.196907525	2.938343663	3.787149685	3.113097503
CASE 10	TFP ($/hr)	702.6553859	697.8020112	695.1486516	707.1955194
	Emission (Ton/hr)	0.393558761	0.617820698	0.570069487	0.323194081
	VD (p.u)	0.53781698	0.451295015	0.476054341	0.539024674
CASE 11	TFP ($/hr)	718.5198728	722.9390367	721.2883434	714.4539224
	Emission (Ton/hr)	0.207167877	0.298024711	0.256604417	0.228137139
	VD (p.u)	0.683290288	0.432628804	0.756848705	0.436191713
	APL (MW)	3.743023082	4.170222863	4.029497391	4.387602145

is systematically recorded in Table 3. Taking a closer look at Case 1, which is woven around three eco-friendly wind power [27] installations, the Total Fuel Price (TFP) touches a figure of 793.998 $/hr. A comprehensive depiction of the TFP convergence trajectory can be observed in Fig. 6. Transitioning to Case 2, it reveals an emission rate for greenhouse gases standing at 0.0923 Ton/hr. Progressing to Case 3, the collective power dissipation (APL) mapped across diverse transmission conduits aggregates to 1.734 MW. In a parallel vein, Case 4's analysis highlights the least voltage fluctuation at each bus point, registering at 0.284 *p.u.* Delving into the Voltage Stability Index, often denoted as the L_max index, note its operational span: from 0 (symbolizing no load) to 1 (representing peak load, verging on voltage instability). Within the ambit of Case 5, the lowermost recorded value for the L_max index settles at 0.134. Navigating to Case 6, the carbon tax metric, abbreviated as C_tax, remains constant at $20 per ton. Factoring in the emission parameters, the all-encompassing financial outlay is calculated to be 823.679 $/hr.

The contrasted outcomes when juxtaposed with three state-of-the-art algorithms accentuate the dexterity and robustness of the proposed MFO technique, especially when grappling with the OPF complexities intertwined with three wind-coal energy units. Switching gears to a multi-objective spectrum, Cases 7 through 9 proactively seek to simultaneously optimize objectives in pairs or trios. The optimization arsenal comprises the innovative MFO framework alongside other cutting-edge techniques like MOGNDO, MOSCA, and MOSMA, mirroring the approach of Scenario 1. The equilibrium solutions, representing the ideal trade-offs, are meticulously curated in Table 4. To visually elucidate the insights, Figs. 7 and 8 graphically represent the multi-objective outcomes for Case 7, Case 8, and Case 9, respectively. Summarizing the exhaustive analysis, the Multi-objective MFO (MOMFO) algorithm unequivocally stands out, positioning itself as a pivotal mechanism for zeroing in on the crème de la crème solutions for the multi-objective OPF conundrum. This is especially true when harmoniously fusing three wind energy units with coal-based power infrastructures.

Table 3 Comparative analysis of single-objective simulations for scenario 2

Case	Control and state variables	MFO	NO	ESO	PIDBSA
Case 1	TFP ($/hr)	793.998	794.433	794.346	796.913
Case 2	Emission (Ton/hr)	0.0923	0.0923	0.0928	0.0923
Case 3	APL (MW)	1.734	1.868	1.794	1.739
Case 4	VD (p.u)	0.284	0.302	0.406	0.353
Case 5	L index	0.134	0.134	0.134	0.134
Case 6	TFP with carbon tax ($/hr)	823.679	824.172	829.143	826.953

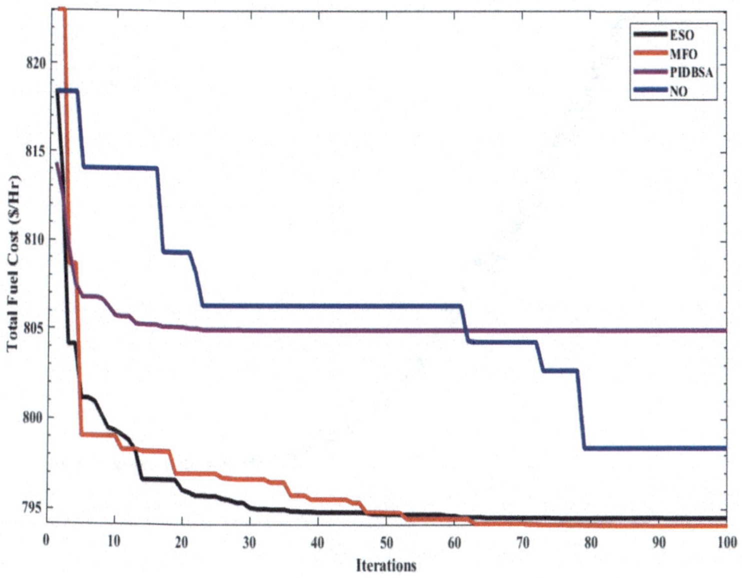

Fig. 6 Convergence of TFP minimization in scenario 2

Table 4 Comparative analysis of multi-objective simulations for scenario 2

Problem	Multi-objectives	MOMFO	MOGNDO	MOSCA	MOSMA
CASE 7	TFP ($/hr)	824.307	825.259	828.710	825.943
	Emission (Ton/hr)	0.418	0.446	0.364	0.402
CASE 8	TFP ($/hr)	854.996	863.525	861.141	862.908
	APL (MW)	3.243	2.973	2.910	2.822
CASE 9	TFP ($/hr)	866.501	865.215	863.884	851.024
	Emission (Ton/hr)	0.113	0.119	0.129	0.202
	APL (MW)	2.945	2.892	3.533	3.322

6.3 *Scenario 3 OPF with Two Wind Power Plants, One Solar Power Plant, and Coal-Based Power Plants*

Each objective, as meticulously detailed in the mathematical structure, undergoes a rigorous evaluation as both individual and multi-dimensional optimization challenges, embracing the latest in metaheuristic methodologies. Techniques such as MFO, NO, PIDBSA, ESO, and OOA are meticulously applied to each objective, ensuring a comprehensive single-objective optimization derived straight from the mathematical formulation. A noteworthy modification at bus stop 13 sees a wind turbine make way for a solar energy module. The control variables adhere to the stipulated constraints consistent with Scenarios 1 and 2. The voltage outputs from

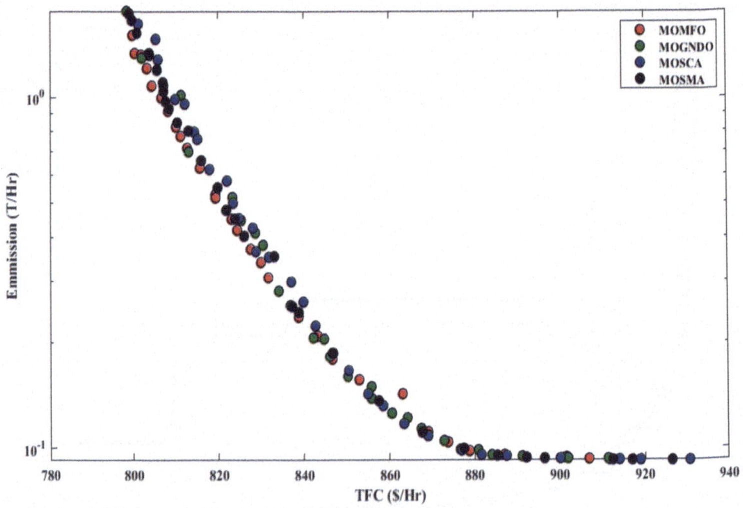

Fig. 7 Pareto front of TFP and emission minimization in scenario 2

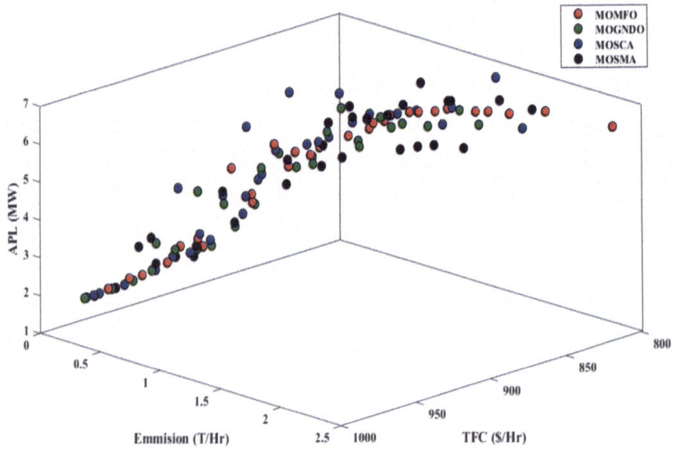

Fig. 8 Pareto front of TFP, emission, and APL minimization in scenario 2

both the turbines and the transformer tap settings conform to the [0.9–1.1] $p.u$ range. It is essential to note that all PQ buses operate within a [0.95–1.1] $p.u$ voltage spectrum, while the reactive power compensator showcases a 0 to 5 MVAr rating.

Table 5 present the optimal lower bounds of the objective functions, inclusive of control variables, spanning from Case 1 through Case 6. Zooming in on Case 1, the combined Total Fuel Price (TFP)—integrating both eco-friendly solar and wind energy infrastructures—arrives at a figure of 780.7627 $/hr. Remarkably, this number trims the margin by 1.7403 $/hr when paralleled with the reference data in [18], which is explicitly highlighted in Table 5. Figure 9 offers a visual representation of the convergence curve for Case 1. The subsequent objective illustrates that the emissions of environmentally detrimental gases peak at 0.092314 tons/hr. An influential factor, the voltage deviation from the standard 1.0 per unit at each bus, greatly impacts the overall grid's operational reliability. Under the umbrella of the third objective, the Aggregate Power Loss (APL) across the intricate network of transmission lines aggregates to 1.735149 MW. Furthermore, Case 4 presents a minimal voltage fluctuation pinpointed at 0.29525 *p.u.* Delving into the world of the Voltage Stability Index—or the L_max index—the range fluctuates between 0 (representing a state of zero load) and 1 (a scenario of peak load, potentially leading to voltage breakdown). In this landscape, Case 5 documents the *L*_max index's floor value as 0.134. In Case 6, the carbon taxation metric (*C*_tax) stands firmly at 20 $/ Ton, as corroborated by the insights in [18]. The cumulative financial implication, when integrated with the emission-centric simulation results, settles at 809.6405 $/hr, presenting an edge over the benchmark set by reference [18], as annotated in Table 5. The synthesized simulation outputs accentuate the robust and superior efficacy of the Moth Flame Optimization methodology in play.

Diving into the realm of multi-objective optimization, the Non-Dominant Sorting (NDS) methodology frequently emerges as a go-to strategy for seamlessly managing myriad objectives in tandem. For the IEEE-30 bus system, a suite of 30 non-dominant solutions is retained to meticulously map the Pareto frontier. Cases 7 through 10 encapsulate the multi-faceted optimization challenges ingrained within the IEEE-30 bus system, harmoniously integrating both wind and solar energy modules. The fuzzy decision-making paradigm takes center stage, diligently discerning the Best Compromise Solution (BCS) amidst the vast expanse of Pareto repositories. The pinnacle of compromise solutions, adorned in bold, finds a home in Table 6. As illustrated vividly in Fig. 10, the Moth Flame Optimization technique emerges as the torchbearer in deriving these solutions. Figure 11 paints a detailed picture of the Pareto frontier for Case 10. Conclusively, the Multi-objective Moth Flame Optimization

Table 5 Comparative analysis of single-objective simulations for scenario 3

Case	Control and state variables	MFO	NO	PIDBSA	ESO	OOA	SHADE Biswas et al. [18]
Case 1	TFP ($/hr)	780.7627	786.1276	793.983	781.241	783.5995	782.503
Case 2	Emission (Ton/hr)	0.092314	0.092321	0.092698	0.092314	0.092403	NA
Case 3	APL (MW)	1.735149	1.744515	1.807462	1.766434	1.759493	NA
Case 4	VD (p.u)	0.290525	0.371194	0.399506	0.319629	0.391915	NA
Case 5	L index	0.134	0.134	0.134	0.134	0.134	NA
Case 6	TFP with carbon tax ($/hr)	809.6405	813.2019	811.6482	809.9711	812.1718	810.346

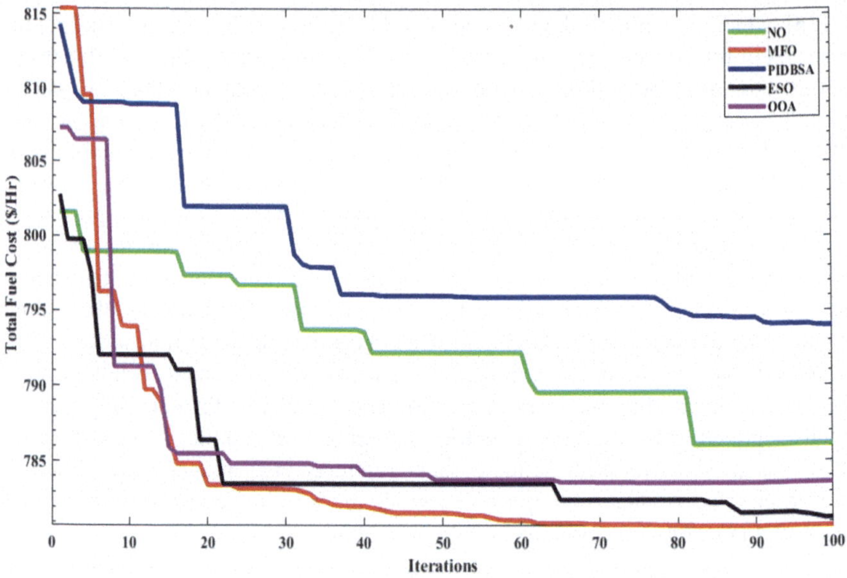

Fig. 9 TFP minimization in scenario 3

(MOMFO) technique solidifies its stature as a premier tool, adeptly navigating and pinpointing optimal resolutions for the multi-objective OPF challenges interwoven with solar and wind energy architectures.

Table 6 Comparative analysis of multi-objective simulations for scenario 3

Problem	Multi-objectives	MOMFO	MOGNDO	MOSCA	MOSMA
CASE 7	TFP ($/hr)	811.078	813.9134	813.9119	816.2233
	Emission (Ton/hr)	0.403312	0.341961	0.379285	0.329001
CASE 8	TFP ($/hr)	838.1847	843.1956	843.0763	842.3003
	APL (MW)	3.209344	3.10038	3.066036	3.20075
CASE 9	TFP ($/hr)	848.8347	849.8022	851.8257	851.15
	Emission (Ton/hr)	0.117143	0.11113	0.129275	0.111584
	APL (MW)	3.463903	3.312307	3.063992	3.110463
CASE 10	TFP ($/hr)	793.7998	796.6186	807.1366	825.2245
	Emission (Ton/hr)	1.185383	1.142331	0.782771	0.699844
	VD (p.u)	0.599355	0.534385	0.521406	0.415817

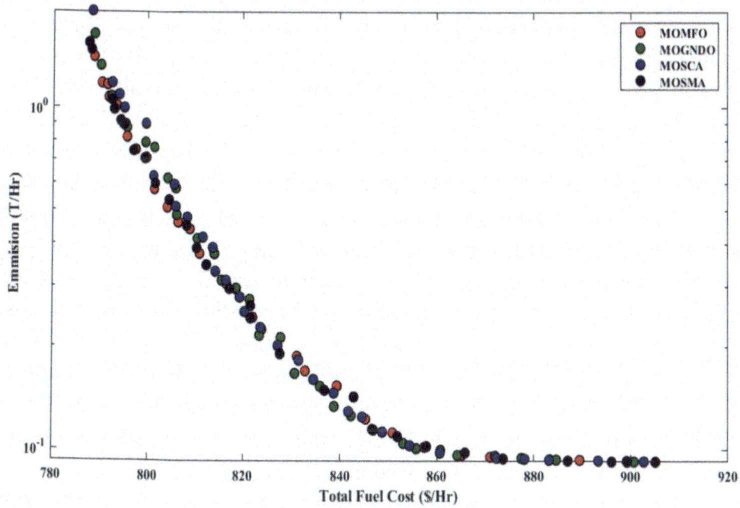

Fig. 10 Pareto front of TFP and emission minimization in scenario 3

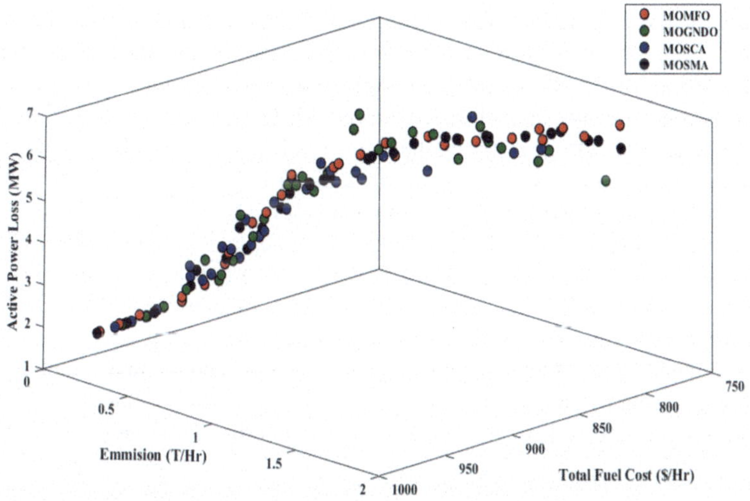

Fig. 11 Pareto front of TFP, emission, and APL minimization in scenario 3

6.4 Scenario 4 OPF with Hybrid Wind-Solar-Small Hydro Power Plants and Coal-Based Power Plants

This research studies the complexities of the stochastic optimal power flow (OPF) challenge, integrating the dynamic trio of wind, solar and miniature hydro power plants. Our approach, grounded in both singular and multi-faceted objective optimization, harnesses the power of the Moth Flame Optimization (MFO) algorithm. In a bid to ensure holistic optimization, the study taps into a robust toolkit of techniques such as the Moth Flame Optimizer (MFO), Nutcracker Optimization (NO), PID-based search Algorithm (PIDBSA), Enhanced Snake Optimizer (ESO), and Osprey Optimization Algorithm (OOA). Diligently curated results, spanning the objective functions and encompassing the minimal values deduced from the latest quartet of algorithms from Case 1 to Case 5, are showcased across Table 7. Notably, when applying MFO, the Total Fuel Price (TFP)—harmonizing the operations of sustainable power plants—peaks at an impressive 891.7198 $/hr. This value stands as a testament to the prowess of MFO, outshining the other algorithms. In a comparative light, it marks a decrement by 1.2832 $/hr against the MOEA/D-SF algorithm and further trims by 1.7832 $/hr relative to the Differential Evolution (MODE) technique (Table 7). Figures 12 and 13 vividly chart the convergence pathways for pivotal parameters such as TFP and Voltage Deviation (VD).

Delving further, leveraging the MFO algorithm, it is observed that the emission quotient of harmful gases in Case 2 reaches 0.092314 tons/hr. Emphasizing grid stability, it is evident that the voltage variance at each bus from the standardized 1.0 per unit offers profound implications. Within the purview of Case 4, the most diminutive voltage fluctuation realized through the moth flame mechanism stands at 0.286557 *p.u*, a testament to its superior performance amidst its contemporaries. When exploring Case 3, the Aggregate Power Loss (APL) imprinted across a myriad of transmission pathways, as deciphered through MFO, stands at 1.86597 MW. Navigating the terrain of voltage stability, the L_max index, serves as a barometer, vacillating between the extremes of 0 (a no-load scenario) and 1 (signifying peak load or a potential voltage meltdown). Venturing into Case 5, the L_max index unveils a floor value of 0.1331. Figures 14 and 15 extend visual comparative landscapes, mapping out metrics like TFP decrement and the downsizing of detrimental gas emissions. When juxtaposing simulation outcomes with contemporaneous algorithms, including but not limited to Nutcracker Optimization (NO), PIDBSA, Enhanced Snake Optimizer (ESO) and Osprey Optimization Algorithm (OOA) and the myriad other referenced optimization mechanics, it is unequivocal that the Moth Flame Optimization methodology propels itself to the forefront with unparalleled efficacy.

Table 7 Comparative analysis of single-objective simulations for scenario 4

Objectives	MFO	NO	PIDBSA	ESO	OOA	MOEA/D-SF	SMODE/SF
TFP ($/h)	891.7198	894.6086	895.8837	891.9316	892.9868	893.003	893.503
Emission (T/h)	0.092314	0.092319	0.093419	0.092314	0.092315	0.1091	0.0961
V.D (p.u)	0.286557	0.304818	0.299236	0.300912	0.30584	NA	NA
Ploss (MW)	1.86597	1.942319	1.891035	1.883791	1.889873	NA	NA
L-max	0.1331	0.1331	0.1331	0.1331	0.1331	NA	NA

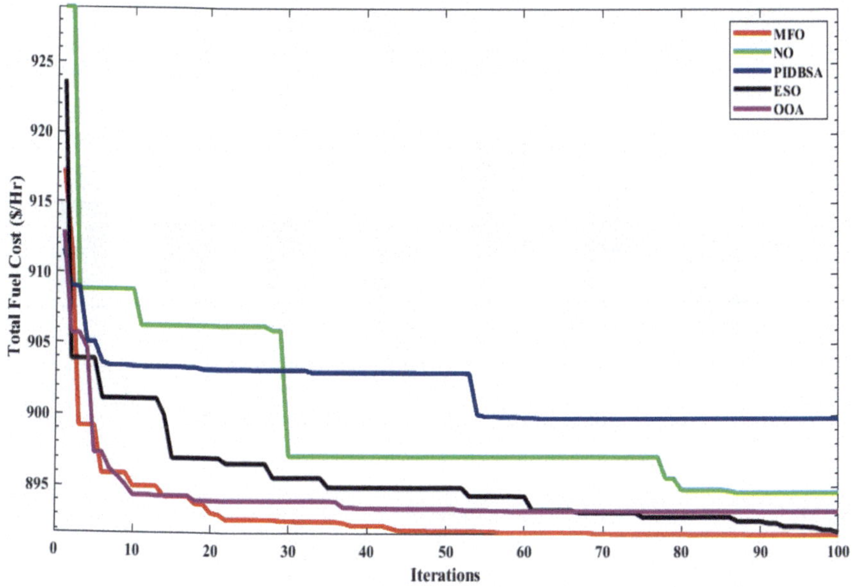

Fig. 12 Convergence of TFP minimization in scenario 4

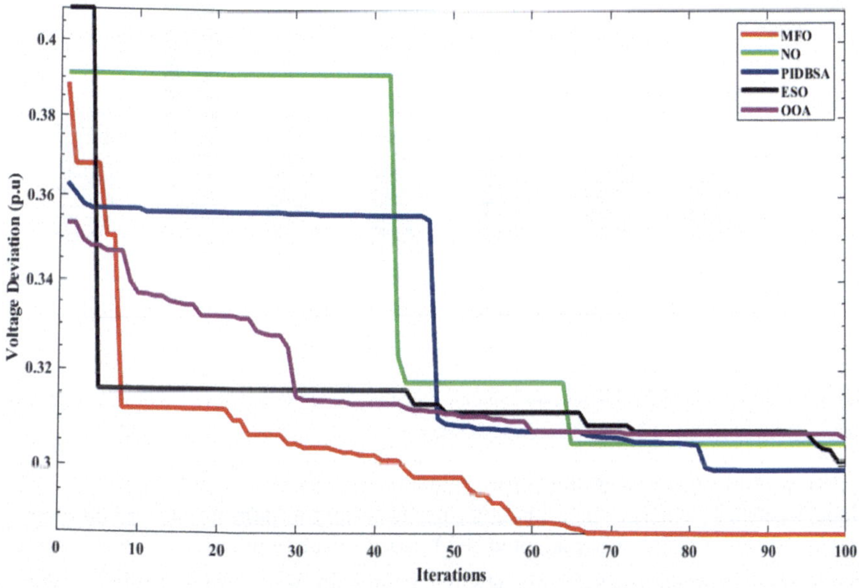

Fig. 13 Convergence of VD minimization in scenario 4

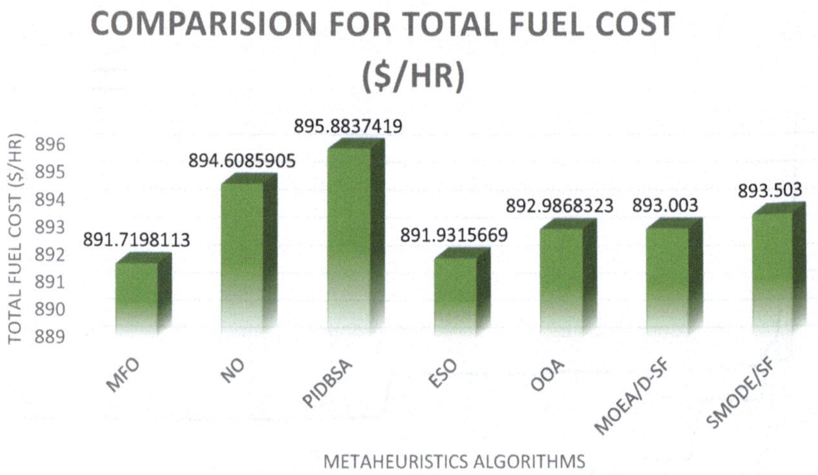

Fig. 14 Comparison of TFP minimization strategies

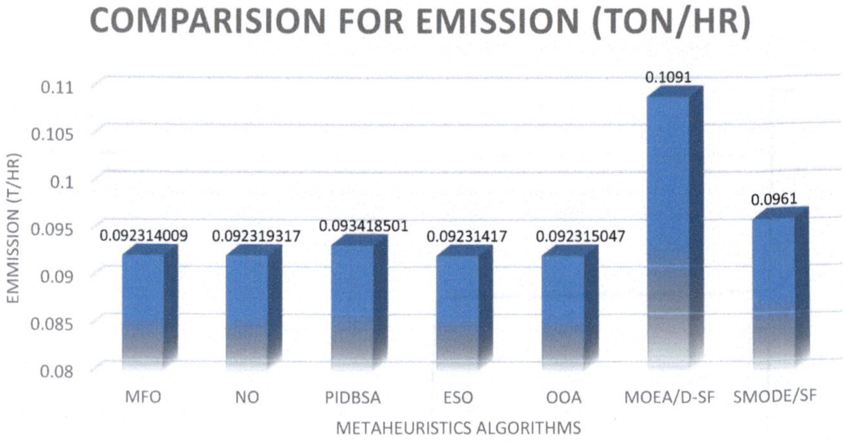

Fig. 15 Comparison of emission minimization strategies

In the multi-objective optimization realm, the Non-Dominant Sorting (NDS) technique frequently emerges as a linchpin, orchestrating a harmonious amalgamation of diverse objectives. For our adapted IEEE-30 bus framework, a bouquet of 30 non-dominant solutions is meticulously retained, paving the path to deciphering the Pareto frontier. Cases 6 to 8 immerse themselves as dual-objective optimization enigmas, while Case 9 evolves as a tri-objective optimization conundrum. Harnessing the fuzzy decision-making prism, the endeavor is to spotlight the Best Compromise Solution (BCS) from the expansive Pareto repositories. The creme de la creme of BCSs, originating from both the MFO algorithm and an array of acclaimed metaheuristic algorithms, take center stage, emboldened in Table 8. Figures 16, 17, 18 encapsulate the essence of Pareto frontiers across Cases 6–9, as shaped by the MFO technique. The proposed methodology locks horns with the celebrated MODE approach [18], as illuminated in Table 9. Synthesizing findings, it is unequivocal that the Multi-objective Moth Flame Optimization (MOMFO) strategy carves a niche for itself, reigning supreme in sculpting optimal resolutions for both singular and multi-dimensional OPF challenges, seamlessly integrating wind, solar and compact hydro power infrastructures.

Table 8 Comparative analysis of multi-objective simulations for scenario 4

Problem	Multi-objectives	MOMFO	MOGNDO	MOSCA	MOSMA
CASE 6	TFP ($/hr)	929.0780871	936.8077994	933.824079	928.8892361
	Emission (Ton/hr)	0.437161464	0.3475164	0.391118272	0.469764347
CASE 7	TFP ($/hr)	978.1408688	975.4980083	974.4323538	971.8752499
	APL (MW)	3.141355785	3.178903677	3.184964952	3.28218407
CASE 8	TFP ($/hr)	900.5077551	903.4519549	899.5682871	904.4630411
	VD (p.u)	0.349915611	0.352237661	0.373742599	0.361428151
CASE 9	TFP ($/hr)	975.1676916	967.3717925	973.7794233	971.1399682
	Emission (Ton/hr)	0.139013965	0.17652361	0.209413865	0.177558537
	APL (MW)	3.347826321	3.541494797	3.662043383	3.375131725

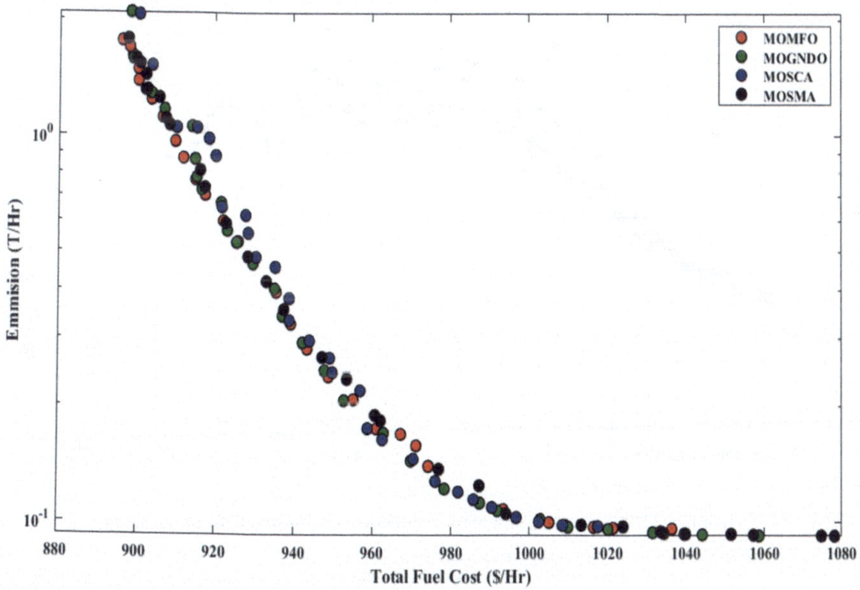

Fig. 16 Pareto front of TFP and emission minimization in scenario 4

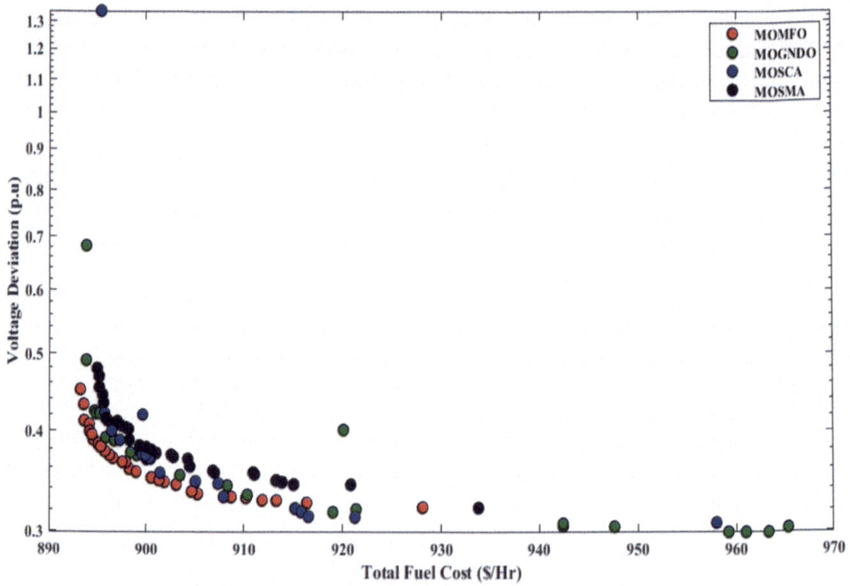

Fig. 17 Pareto front of TFP and VD minimization in scenario 4

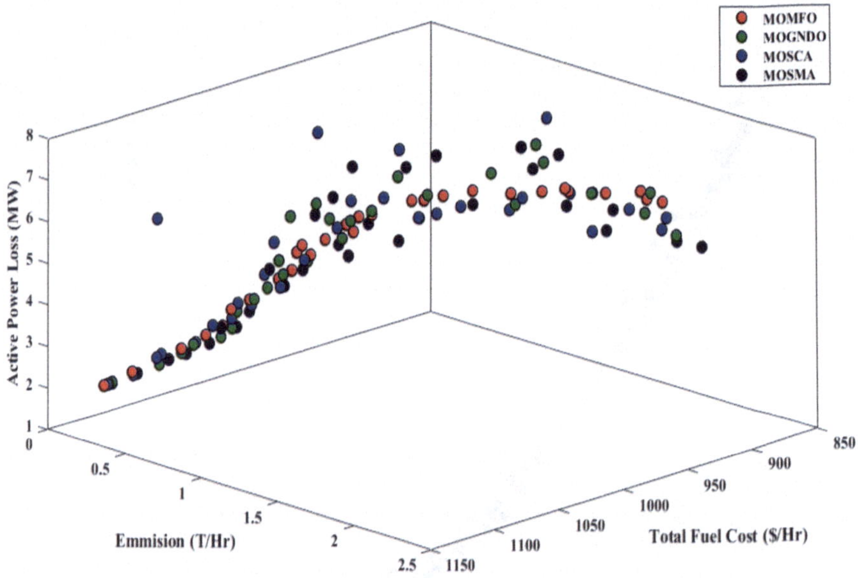

Fig. 18 Pareto front of TFP, emission, and active power loss minimization in scenario 4

Table 9 Comparative analysis of multi-objective simulations for case 6

Multi-objectives functions	MOMFO	SMODE/SF	MOEA/D-SF
TFP ($/h)	929.0780871	927.049	919.040
Emission (T/h)	0.437161464	0.4721	0.6221

Moth Flame Optimization (MFO) algorithm has been rigorously tested across various scenarios and compared with several recent optimization techniques. The MFO method proves to be adaptable and efficient, addressing both single-objective and multi-objective OPF problems across various scenarios integrating different renewable energy resources. MFO outperforms recently developed optimization algorithms, such as Nutcracker Optimization (NO), Enhanced Snake Optimizer (ESO), PID-based search Algorithm (PIDBSA), and Osprey Optimization Algorithm (OOA) in multiple scenarios. 3. Across scenarios, the MFO technique often achieves reduced Total Fuel Price (TFP) and overall costs. In Scenario 4, for instance, TFP with MFO, inclusive of sustainable power plants, demonstrates the best result among the four algorithms and registers significant savings when compared with other methods. The pollutant gas emission rates achieved using MFO in the tested scenarios remained consistent and often surpassed the performance of other optimization techniques. This emphasizes the environmental benefits of using the MFO method. Voltage deviation, Aggregate Power Loss (APL), and the L-max index are critical indicators of grid stability and reliability. In various scenarios, the MFO algorithm yielded results ensuring improved grid performance. The study establishes that when dealing with multi-objective optimization problems, especially in renewable-integrated energy systems, the MFO algorithm finds the best compromise solutions, striking a balance among competing objectives.

In the scenario that integrated wind, solar, and small hydro power plants with traditional coal-based plants, the MFO method stood out as a leading optimization technique. This highlights its potential in a diversified energy landscape. The proposed MFO approach is executed on standard hardware, highlighting its efficiency and potential for real-world application without requiring supercomputing resources. Thus, the Moth Flame Optimization (MFO) method emerges as a robust, efficient and promising tool for addressing the challenges of Optimum Power Flow in systems integrating Renewable Energy Resources. This research substantiates the capability of MFO to optimize both economic and environmental objectives in energy systems, pushing the frontier for sustainable energy solutions. As the energy sector moves toward a more sustainable and integrated model, techniques like MFO will be paramount in ensuring optimal resource allocation and system performance.

7 Conclusion

The development and implementation of the Multi-objective Moth Flame Optimization (MOMFO) technique for addressing single and multi-objective Optimum Power Flow (OPF) challenges represents a significant advancement in the integration and management of Renewable Energy Resources (RER) within power systems. Scientifically, this research contributes to the field by presenting a novel optimization approach that not only showcases superior convergence performance over existing algorithms but also enhances the feasibility of integrating diverse renewable energy sources into the grid. The MOMFO technique's practicality and efficiency in handling single-objective as well as multi-objective OPF problems demonstrates its transformative potential as a power system optimization tool.

Socially, this research is far-reaching. By integrating renewable energy sources like solar, wind and small hydroelectric plants efficiently, the MOMFO approach supports the movement to a more sustainable and environmentally friendly energy landscape. This initiative is very important in terms of global climate change issues since it leads to reduction of greenhouse gas emissions and dependency on fossil fuels; thus, it contributes toward environmental conservation and energy security. Moreover, effective optimization tools like MOMFO can drive up adoption rates for clean energy solutions which will have economic benefits such as reduced energy costs while also creating green jobs hence meeting some of the sustainable development goals hence should be encouraged.

This MFO technique suggestion poses scientific research-based questions for future studies in power system optimization that are significant also for society because they promote use of renewable energy resources. Such research advances resilience in power infrastructures against optimizing power flow under new mixtures of traditional and modern forms of fuels themselves including renewables. Therefore, there is need for further investigations to enhance performance using these building blocks, with an aim of developing enhanced MFO techniques which can accommodate ever increasing complexities associated with present day energies.

The reason behind this investigation is both scientific and social since it has potential to impact global power systems' efficiency and sustainability significantly. This work contributes to wider objectives by enhancing optimization techniques that encourage integration of renewable energy resources for purposes of carbon emission reductions as well as improving countries' abilities to adapt economically during difficulties thus addressing current woes faced by humanity today.

References

1. Carpentier JL (1985) Optimal power flows: uses, methods and developments. IFAC Proceed Vol 18:11–21. https://doi.org/10.1016/S1474-6670(17)60410-5
2. Wolpert DH, Macready WG (1997) No free lunch theorems for optimization. IEEE Trans Evol Comput 1:67–82. https://doi.org/10.1109/4235.585893

3. Ming NIU, Can WAN, Zhao XU (2014) A review on applications of heuristic optimization algorithms for optimal power flow in modern power systems. J Modern Power Syst Clean Energy 2:289–297. https://doi.org/10.1007/s40565-014-0089-4
4. Roy R, Jadhav HT (2016) Optimal power flow solution of power system incorporating stochastic wind power using Gbest guided artificial bee colony algorithm. Int J Electr Power Energy Syst 64:562–578. https://doi.org/10.1016/j.ijepes.2014.07.010
5. Panda A, Tripathy M (2014) Optimal power flow solution of wind integrated power system using modified bacteria foraging algorithm. Int J Electr Power Energy Syst 54:306–314. https://doi.org/10.1016/j.ijepes.2013.07.018
6. Panda A, Tripathy M (2015) Security constrained optimal power flow solution of wind-thermal generation system using modified bacteria foraging algorithm. Energy 93:816–827. https://doi.org/10.1016/j.energy.2015.09.083
7. Jabr RA, Pal BC (2009) Intermittent wind generation in optimal power flow dispatching. IET Gener Trans Distrib 3(1):66–74. https://doi.org/10.1049/iet-gtd:20080273
8. Mishra S, Mishra Y, Vignesh S (2011) Security constrained economic dispatch considering wind energy conversion systems, In: 2011 IEEE Power and energy society general meeting, https://doi.org/10.1109/PES.2011.6039544
9. Zhou W, Peng Y, Sun H (2010) Optimal wind-thermal coordination dispatch based on risk reserve constraints. Eur Trans Elect Power 21:740–756. https://doi.org/10.1002/etep.474
10. Dubey HM, Pandit M, Panigrahi BK (2015) Hybrid flower pollination algorithm with time-varying fuzzy selection mechanism for wind integrated multi-objective dynamic economic dispatch. Renew Energy 83:188–202. https://doi.org/10.1016/j.renene.2015.04.034
11. Tazvinga H, Zhu B, Xia X (2015) Optimal power flow management for distributed energy resources with batteries. Energy Convers Manage 102:104–110. https://doi.org/10.1016/j.enconman.2015.01.015
12. Kusakana K (2016) Optimal scheduling for distributed hybrid system with pumped hydro storage. Energy Convers Manage 111:253–260. https://doi.org/10.1016/j.enconman.2015.12.081
13. Reddy SS, Bijwe PR, Abhyankar AR (2015) Real-time economic dispatch considering renewable power generation variability and uncertainty over scheduling period. IEEE Syst J 9:1440–1451. https://doi.org/10.1109/JSYST.2014.2325967
14. Salkuti SR (2019) Day-ahead thermal and renewable power generation scheduling considering uncertainty. Renew Energy 131:956–965. https://doi.org/10.1016/j.renene.2018.07.106
15. Salkuti SR, Sandeep V, Babu BC, Jung CM (2018) Multi-objective based optimal generation scheduling considering wind and solar energy systems. Int J Emerg Elect Power Syst. https://doi.org/10.1515/ijeeps-2018-0006
16. Pandya SB, Visumathi J, Mahdal M, Mahanta TK, Jangir P (2022) A novel MOGNDO algorithm for security-constrained optimal power flow problems. Electronics 11:3825. https://doi.org/10.3390/electronics11223825
17. Pandya S, Jariwala HR (2021) Single—and multiobjective optimal power flow with stochastic wind and solar power plants using moth flame optimization algorithm. Smart Sci 10:77–117. https://doi.org/10.1080/23080477.2021.1964692
18. Biswas PP, Suganthan PN, Qu BY, Amaratunga GAJ (2018) Multiobjective economic-environmental power dispatch with stochastic wind-solar-small hydro power. Energy 150:1039–1057. https://doi.org/10.1016/j.energy.2018.03.002
19. Mirjalili S (2015) Moth-flame optimization algorithm: a novel nature-inspired heuristic paradigm. Knowl-Based Syst 89:228–249. https://doi.org/10.1016/j.knosys.2015.07.006
20. Savsani V, Tawhid MA (2017) Non-dominated sorting moth flame optimization (NS-MFO) for multi-objective problems. Eng Appl Artif Intell 63:20–32. https://doi.org/10.1016/j.engappai.2017.04.018
21. Yao L, Yuan P, Tsai C-Y, Zhang T, Lu Y, Ding S (2023) ESO: an enhanced snake optimizer for real-world engineering problems. Expert Syst Appl 230:120594. https://doi.org/10.1016/j.eswa.2023.120594

22. Gao Y (2023) PID-based search algorithm: a novel metaheuristic algorithm based on PID algorithm. Expert Syst Appl 232:120886. https://doi.org/10.1016/j.eswa.2023.120886
23. Dehghani M, Trojovský P (2023) Osprey optimization algorithm: a new bio-inspired metaheuristic algorithm for solving engineering optimization problems. Front Mech Eng. https://doi.org/10.3389/fmech.2022.1126450
24. Tawhid MA, Savsani V (2019) Multi-objective sine-cosine algorithm (MO-SCA) for multi-objective engineering design problems. Neural Comput Appl 31:915–929. https://doi.org/10.1007/s00521-017-3049-x
25. Premkumar M, Jangir P, Sowmya R, Alhelou HH, Heidari AA, Chen H (2020) MOSMA: multi-objective slime mould algorithm based on elitist non-dominated sorting. IEEE Access 9:3229–3248. https://doi.org/10.1109/ACCESS.2020.3047936
26. Saka K, Adebanji B, Olulope P, Fasina T, Abe A, Ajeba W (2022) Modeling and simulation of small hydro-solar PV hybrid generating system for complementary power supply in a metropolitan city. Appl Eng Lett 7(4):172–180. https://doi.org/10.18485/aeletters.2022.7.4.5
27. Koondhar MA, Ali M, Keerio MU, Junejo AK, Laghari IA, Chandio S (2021) Wind energy conversion system using maximum power point tracking technique—a comprehensive survey. Appl Eng Lett 6(4):148–156

Improving Primary Storage Performance Using Data Deduplication and Pre-fetching Methods in Cloud

M. Bhavya, Buddesab, J. Thriveni, and K. R. Venugopal

Abstract The Cloud Storage Server (CSS) provide huge storage facilities to many data users. The clients outsource large amount of data to the CSS. One of the challenging issues to CSS is data redundancy. Many researchers have proposed advanced techniques to eliminate data redundancy, in which data deduplication is one of the techniques to eliminate duplicate data in the cloud. To improve the primary memory management, the authors have introduced the Performance-Oriented I/O Deduplication (POD) in the existing scheme. The existing scheme has read amplification problem and consumes more network bandwidth while downloading a file. To overcome these issues, in this chapter, a data Pre-fetching scheme is proposed with primary level data deduplication. The proposed scheme uses the linear regression technique and data Pre-fetching together to analyze the I/O disk access with client piggybacking information. The proposed scheme is useful to clients with less computing resources and limited power. The objective of the proposed work is to reduce the client data download time and reduce the data access process. The performance analysis shows that the proposed scheme improves storage server access efficiency by resolving read amplification problem and minimizes network latency, network traffic and energy consumption as compared to the existing scheme.

Keywords Cloud storage server · Data deduplication · Data Pre-fetching · Linear regression · Client file system · Piggybacking

M. Bhavya
Department of AI & ML, M S Ramaiah Institute of Technology, Bangalore 560054, India

Buddesab (✉)
Department of Engineering, Dr APJ Abdul Kalam School of Engineering, Bangalore, KAR 560049, India
e-mail: buddesab@gcu.edu.in

J. Thriveni
University of Visvesvaraya College of Engineering, Bangalore University, Bangaluru, KAR 560001, India
e-mail: drgowda@gamail.com

K. R. Venugopal
Bangalore University, Bangaluru, KAR 560001, India

A. Kumar et al. (eds.), *Trends in Mechatronics Systems*, Emerging Trends in Mechatronics, https://doi.org/10.1007/978-981-97-9108-8_12

1 Introduction

In present days we can see many industrial sectors opting for cloud storage to reduce their computing infrastructure overhead. The employees are utilizing cloud resources to perform different types of operations based on their requirements. Sometimes the similar data is stored by different employees leads to redundant data. For example, consider in companies administration sends a file with 10 MB size to all employees through email. In case all employees backup the file on the same storage system, then multiple copies of the same file will be stored in the storage server. This will lead to data deduplication.

A specific method called deduplication is used to remove the data redundancy in the CSS and makes sure that a single copy of data is stored in the server. The execution of deduplication in a cloud server saves the storage utilization cost, storage space consumption, and eliminates data redundancy. The deduplication techniques are at different levels: block-level, file level, source level, and destination level. Based on users requirement they select different deduplication techniques and fulfill their needs.

The recent analysis shows that increase in data size and its volatile characteristic, it is a challenging task for a storage system [1]. It causes input/output bottlenecks and degrades the primary storage system performance. The Offline dedup [2] and iDedup [3] are the data deduplication schemes used to achieve the better primary storage system performance. But they are not considering the files with small size, which leads to conjunction in communication. To overcome that Mao et al. [4] have proposed Performance-Oriented I/O Deduplication (POD) for managing primary memory by implementing iCache and Selective dedup schemes. These schemes reduce network traffic while writing and transferring data in the storage system. To reduce file downloading time-based on the user read request and increase the system throughput we proposed a scheme that stores the frequently accessed files in a client file system. If read request sent to access a file, the read request hits to the client file system, the client machine checks for the data availability in buffered data and respond to the client, else request is forwarded to the CSS. This method reduces the file download time, network bandwidth usage, and energy consumption especially for those files which are accessed very frequently.

1.1 Motivation

The existing scheme implemented data deduplication over small data files to perform primary memory management. The authors try to adjust the read and write cache based on the file size to perform file upload/download. But the complete user data is stored in the CSS. Every time users need to request and access data from CSS, which leads to network traffic, takes extra time to download a file, and consumes more energy.

1.2 Research Contribution

In this chapter, we improved storage efficiency by implementing a scheme called Data Pre-fetching. The client file system stores the pre-fetched data in its buffer, in which we store frequently accessed files for some time. Therefore, a user can access the file faster and check for duplication without uploading it to the cloud server, this reduces the network traffic, file download time, and system utilization.

1.3 Objectives

The main objectives of the proposed scheme are:

1. Reduce read response time.
2. Minimize the energy consumption.
3. Efficient use of storage.
4. Increase the throughput.

1.4 Chapter Organization

The chapter is organized as follows. Section 2 highlights the related work that is used for the implementation. Section 3 explains the existing work. Section 4 presence the system model which describes the modules used in the proposed scheme and presents the system design, Sect. 5 presence the performance evaluation of the existing and proposed scheme with results. Finally, Sect. 6 contains the conclusions.

2 Related Work

Xia et al. [5] have proposed a similarity-locality based deduplication system (SiLo) that makes use of both locality and similarity in backup streams to eliminate duplicates at an extremely lower RAM overhead and achieve higher deduplication throughput. The similarity and locality approach are implemented in existing scheme individually, that works very slow for large data and not suitable for higher deduplication performance. In the proposed scheme authors combine both similarity and locality-based approaches. The proposed scheme performs deduplication efficiently, reduces RAM usage and provide higher throughput. The SiLo indexing scheme is not suitable for the deduplication applications in primary and cloud storage.

Mao et al. [6] have introduced SAR, Solid-state drives (SSDs) Assisted Restore scheme. The data deduplication process with unique data sharing characteristic implemented in SSDs-based storage system. That efficiently performs the high random read operation and low power consumption. The deduplication process results to the unique data blocks of small size with high reference count. Data deduplication causes fragmentation which can result in following read function need to request HDD I/O access multiple times, to resolve that SAR is implemented. The experimental results show that SAR execute the traditional deduplication-based schemes vitally and help to restore data and reduce the energy consumption efficiently.

To assess the performance of deduplication and chunking aspects, El-Shimi et al. [7] have implemented a new deduplication engine. The existing system which used regular indexing with or without data partition consumes high cache or index memory usage. The deduplication engine executes deduplication technique by producing a uniform sized chunk distribution, reducing the metadata storage utilization. Reduction of index memory usage by using optimized indexing and data partition achieve high deduplication savings at low computational overhead. Other important aspects of primary data storage, reliability and resiliency are not considered.

To optimize the storage space utilization, Koller et al. [8] have introduced an I/O Deduplication that performs content similarity check by eliminating I/O operations for improving I/O performance. The existing scheme eliminates duplicate data from the cloud storage system to save storage space. Authors have proposed I/O Deduplication for improving performance that consists of three advance techniques: selective deduplication, content-based caching, and dynamic replica retrieval. Significant overall performance improvement is observed in the system. Write I/O operation is not considered for content-based optimizations in the proposed work.

To improve the random disk I/O performance in file systems, Schindler et al. [9] have designed proximal I/O, a new technique. The existing update-in-place or write-anywhere methods are used for serial read and write access but it translates into inefficient non-sequential I/O. By using flash memory-based write cache, the proximal I/O accumulates random updates till it reaches sufficient capacity to make full use of proximal I/O. The results shows that, it is possible to service 5.3 user I/O requests per revolution with a cache of just 1% of the overall disk-base storage capacity.

Srinivasan et al. [10] have proposed inline deduplication, iDedup for primary workloads while minimizing extra I/Os and seeks. The primary workloads due to introduction of extra latency and deduplicating causes fragmentation which increases seeks for data. iDedup is based on real-world workloads which selective deduplicate data, thereby retaining metadata in the cache, iDedup efficiently sidesteps fragmentation and capitalizes on temporal locality, mitigating additional I/Os. It delivers substantial deduplication gains while imposing minimal resource overheads in terms of CPU and memory.

Hua et al. [11] have proposed a scheme which is less complex and easy to implement for Avionics Full DupleX (AFDX) networks. This scheme is called Deduplication-aware Deficit Round Robin (D2DRR). The existing analysis focuses on end-to-end delays with FIFO and static priority scheduling. D2DRR is capable of handling dynamic workloads exhibited by many avionic flows and flexible enough to handle burst flows. Bandwidth saving is achieved by implementing data deduplication.

Meyer et al. [12] have analyzed the relative efficiency of data deduplication by eliminating redundancy at file level and block-level. They studied file fragmentation in a commercial environment on nearly one thousand Windows file systems. The efficiency of the existing methods of deduplication is low with a low cost and high fragmentation rate. Fragmentation is manageable when file deduplication with sparseness efficiently lowers storage consumption. But the proposed method is not supported for namespace and unstructured files [21].

A History Aware Rewriting (HAR) algorithm have introduced by Fu et al. [13], which accurately identifies and rewrites fragmented chunks based on historical data of backup systems. The proposed Container-Marker Algorithm (CMA) is used to identify valid containers. The restore performance is improved by rewriting duplicated and fragmented chunks. The invalid chunks are reclaimed by identifying and merging fragmented chunks into new containers. Using data from real world, the results show that the proposed system improves the restore performance and reduces the metadata overhead.

To improve online RAID reconstruction performance, Wu et al. [14] proposed Intelligent Data Outsourcing (IDO) which is a proactive and zone-based optimization technique. The existing approach fails to speedup data recovery from disk failure and high online RAID reconstruction performance. The IDO addresses the drawbacks of the existing system. Extensive trace-driven experiments show that IDO simultaneously speeds up the reconstruction time and the average user response time and improves performance [22].

3 Background Work

In this section, we have explained the POD approach which consists of two methods called select Dedup and iCache to improve primary memory performance.

3.1 iCache

As we know that in primary storage the input/output workload may keep changing intermittently due to read and write operations. The iCache [15] is designed to improve primary storage performance. iCache manages the ghost read cache and ghost index cache which contains metadata of actual data stored in secondary storage. The flushed-out data from the read and index cache will be stored in the corresponding ghost cache. These methods help to increase the cache size based on the I/O load and helps predict the cost–benefit of the actual cache. Figure 1. Shows the cache deduplication in which the read/write operation contains the source address of the request

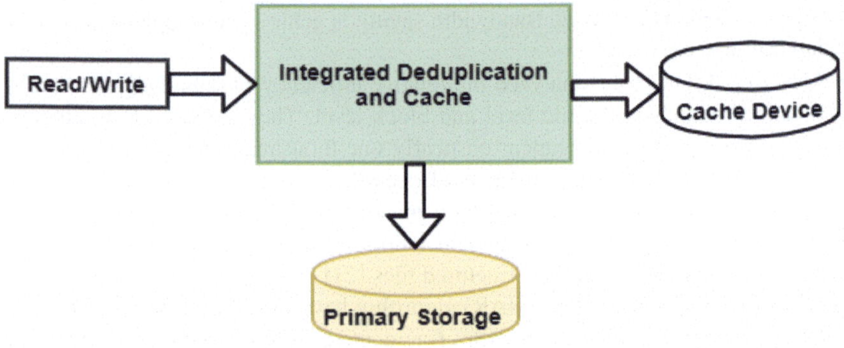

Fig. 1 Cache deduplication

which the user gives as input. The read cache and index cache can be resized dynamically, but the DRAM size is fixed in storage controller. The actual block contains metadata and data cache. Metadata cache maps the address of primary storage to a fingerprint. The cache block contains data blocks. The duplication-aware algorithm is implemented to manage Metadata Cache and Data Cache replacement.

3.2 Select Dedup

The Select Dedup is used during data write operation to reduce the write traffic by updating the Map table if the write request is redundant. The Select Dedup classifies data blocks into three groups based on the number of redundant blocks. If redundant blocks are less than the threshold then those blocks will be uploaded to the backend storage device else the map table needs to be updated for reference. If redundant data is high, it helps to save storage space. The Select Dedup method causes a read amplification problem on these data while reading requests, which reduces the data read performance.

The Index and Map table are the two data structures used in Select Dedup to perform redirection of the I/O requests and deduplication. The Index table stores the encrypted value of the latest data that is uploaded to the storage server. The Map table helps to store complete details of the deduplicated write requests whose write data are already uploaded on the server.

4 Proposed Model

To enhance the existing scheme performance, we have introduced a new technique called data Pre-fetching. Based the disk I/O access history, the CSS can forecast future I/O accesses and authorize physical read requests for data retrieval. This strategy enhances system performance and refines I/O access efficiency.

The proposed architecture contains four entities is shown in Fig. 2.

- **Clients:** The users who access the cloud data are called clients. The clients can send both read and write requests to the files on the cloud. They perform multiple operations based on their requirement. If a write request is sent by the user, then the file is split into blocks and the hash value of each block is compared with the Index table in cache memory. If no value is found, the file is uploaded to cloud storage else Map table is updated.
- **Client File System (CFS):** Client file system executes the user application. The frequently accessed data is pre-fetched and stored in the client file system. The client file system replies to the users read request with the buffered data based on the user access history. This helps to speed up the file downloading time.
- **Storage Node:** The storage node helps to perform primary memory management. POD is present within the storage node and communicate with the File Systems through the standard read/write interface. The two main operations like Select Dedup and iCache are executed in the storage node.
- **Hard Disk Drive (HDD):** The HDD is a storage device used to store and retrieve binary data. Multiple users store their structured or unstructured data. Managing

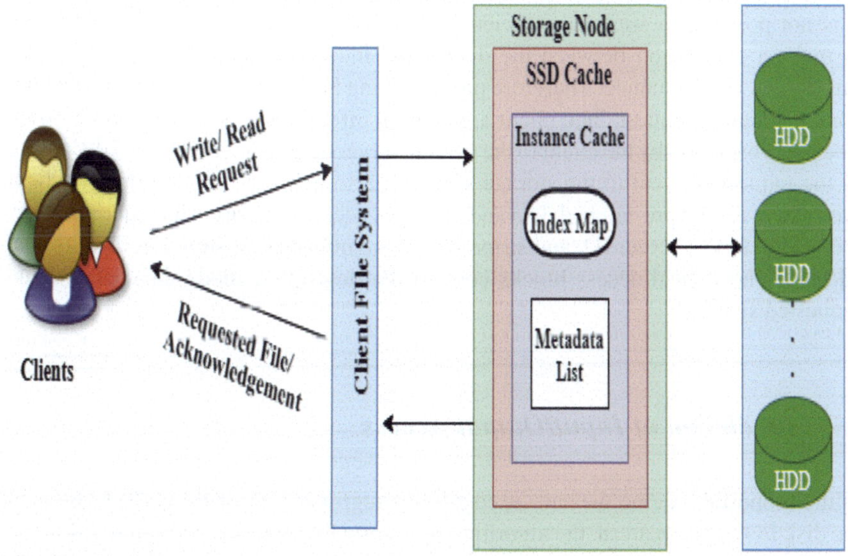

Fig. 2 System model

such huge data is a complex task for HDD. The deduplication concepts help to eliminate the redundant data in HDD and save the storage space.

The general typical is implemented as follows. When the client need to outsource data to the cloud storage, he/she will send a file to the storage node through the Internet. The Select Dedup consists of two modules namely, Request Redirector and Data Deduplication present in the storage node. The Data Deduplication splits the received data into a n blocks and execute the hash function to generate hash values for each block. Using these hash values it will check whether data is redundant or not. The Request Redirector helps to check whether data must be deduplicated or not and helps to reduce the writing overhead. An Access Monitor and Swap Module are present in iCache. The Access Monitor controls the incoming read/write hits rates and intensity. Based on the Access Monitor information, Swap Module modifies the cache space within the read and index cache. To overcome read amplification problem while writing a data, we have introduced data Pre-fetching for the read request. Based on I/O access of the disk, the storage server provides physical request to the user. Once the read request is executed the data is provided to the clients without delay by the client file system [16]. This process helps to reduce the network overhead, reduces the file downloading time and energy consumption.

4.1 Piggybacking

Most of the researchers concentrate on the logical I/O access data which is present in the client file system to form the I/O access pattern. Using logical I/O access it is not possible to create an efficient I/O access pattern. It is a challenging task to make a connection between the distributed file system and the applications to improve I/O performance. Therefore piggybacking [17] concept is introduced which contains logical request data, client application information, and client file system. The piggyback sent by the client to the storage server helps to record the I/O disk event of the requested client to the storage server. It also helps in determining the client file system for the pre-fetched data and allocates access patterns. The requested size, block ID, storage server ID, and stripe ID are the information stored in the physical address. The logical access blocks contain file descriptor, inode data, offset, and requested size.

4.2 Prediction of Input/Output Access

In the proposed scheme we have used a linear regression algorithm [18] to forecast the disk I/O access pattern, the algorithm requires a linear access pattern in the series of present access activities to accurately predict future accesses. The new access pattern contains Addr_{cur} as starting address, Size_{cur} request size, and T_{cur} Current access time. It is labeled as Access ($Addr_{cur}$, $Size_{cur}$, T_{cur}). The future expected access prediction will be having Access ($Addr_{new}$, $Size_{new}$, T_{new}). After a point of time, the proposed algorithm predicts all block access activities which occurred recently from the existing block access history.

Algorithm 1: Linear regression algorithm for access prediction

Step 1: **Start**
Step 2: Count the total number of access in *Access_set*
Step 3: **if** (Count exceed Threshold)

$$SLOPE = Data_{amount}/Time_{used}$$
$$T_{serving} = Size_{cur}/Data\ Throughput$$
$$Addr_{new} = Addr_{cur} + T_{serving}/SLOPE$$

Step 4: **else**
Not a linear sequence goto **end**
Step 5: **end if**
Step 6: Calculating the predicted access pattern address

$$T_{new} = T_{cur} + T_{serving}$$
$$Size_{new} = Size_{cur}\ \text{or}\ Average\ size$$

Step 7: **end**

Algorithm1. Describes the proposed linear regression method for access prediction. In this algorithm, if the required block addresses fall within the given range, then we count the total number of access that occurred. The threshold variable is a predefined value that is depends on the density of I/O operations. If total access count is larger than the threshold value then it ensures that the access request belongs to a linear access sequence. The sequential access pattern is called SLOPE, which is then measured by adding the data size during linear access sequence and time taken for completing all related access requests in the current time period. Likewise, $T_{serving}$ is the amount of time required to execute the currently received read request. If the current access sequence follows a sequential tendency, it is assumed that the future access request is belongs to the linear access sequence. The future request size, address and time are calculated in the last two steps. By this method, we can predict the future access request efficiently.

4.3 Data Pre-fetching

Data Pre-fetching is an efficient scheme to upgrade the I/O access process. In this approach, the cloud storage server predicts the prospective read request by considering the I/O disk history. This facilitates the storage server in proactively issuing physical read requests to retrieve data. The CSS then forwards the Pre-fetched data to the relevant client file system, which remains uninvolved in any data Pre-fetching or prediction processes. It just sends data to the particular client who sends the read request with the buffered data. The proposed scheme reduces read latency on the client-side and also helps the client system with less computing and energy resources.

In the proposed scheme we have used PARTE [19], a prototype parallel file system with three entities. *PARTE-client file system* prepares the piggybacking for I/O request by collecting the information about applications and the client file system. It collects file data and actual metadata details obtained by sending the request to the particular servers separately. Based on the guidelines *PARTE metadata server* provides instruction to add or delete the data to all storage servers. *PARTE storage server* performs access prediction and pattern classification by considering piggybacking information.

5 Results and Performance Valuations

The results of implementation of the existing system and proposed system are compared in the section. The file downloading process from cloud storage or the time required to satisfy a read request method consumes more time, network bandwidth and Energy. To overcome this and to reduce the workload of the client, we have proposed a scheme that stores the frequently accessed files in the client file system based on the I/O disk access. Frequently sending read requests to access a file go to the client file system, if the requested data is available in the buffer, then the client file system responds to the client with the pre-fetched data [20, 21]. If data is not present in the buffer, then the client file system fetches the file from the cloud. This method reduces the download time, read latency, network bandwidth and energy especially for those files which are accessed very frequently.

To check the benefits of the proposed system we have used the institutional test data set, which represents the block traces of online transaction processing applications. In the proposed implementation, based on the read request, the client file system with the size 120 MB issues the I/O request to the storage server for read/write block data, according to log file. The Least Recently Used (LRU) replacement algorithm [22] is used in the client file system to update the buffer with frequently accessed data [23]. The threshold value is set based on the frequency of I/O operations. In the proposed scheme we have set the threshold value 10. Storage server performs access prediction with threshold value and piggybacking [24, 25] information. With all these advance features we were able to achieve the better I/O read response.

5.1 File Downloading Time

The time consumed to download a data using both existing and proposed systems is shown in Table 1. The file size is in MB and the time consumption is calculated

Table 1 File downloading process using data Pre-fetching technique

Sl. No	File size in MB	Existing system without Pre-fetching (Time in ms)	Proposed system with Pre-fetching (Time in ms)
1	5	203.39	162.15
2	10	297.16	233.37
3	15	315.64	284.68
4	20	401.91	337.17
5	25	450.00	412.70
6	30	537.17	480.10

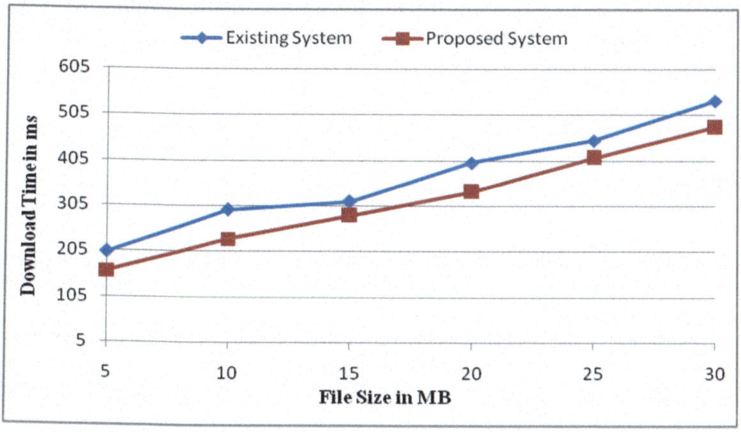

Fig. 3 Comparison of file downloading time in the existing and proposed system

in milliseconds. In the existing system, the file needs to be accessed from storage server every time. This process consumed more time and causes read amplification problem and delay in file downloading. To overcome this problem we proposed a data Pre-fetching method and improved the access efficiency.

The file downloading time is compared with and without the data Pre-fetching technique as shown in Fig. 3. From the graph, we can conclude that the proposed system enhances the system performance by an average of 15 percent compared to the existing system.

6 Conclusions

The proposed system performs primary level data deduplication to eliminate the redundant data in CSS. The data Pre-fetching technique is implemented on storage server. After analyzing the existing log storage server is able to predict the future

I/O access to guide the fetching data in advance and send the Pre-fetched data to the client file system. To design adequate, I/O access pattern and forward exact Pre-fetched data, client node piggybacked its client file system information with relevant I/O access and sends it to the storage server. Based on the client information and the I/O disk access history the CSS will transfer the frequently accessed data to the client file system. Therefore, instead of requesting CSS [24, 25] to read data, the client directly accesses the data from the client file system. In this way, the proposed system reduces the read amplification problem, reduces the file downloading time, network latency, network traffic, and energy consumption [26]. The block-level data storage needs more space to store the Map table. To prevent the Map table data loss from power failure, we used non-volatile memory. This consumes more memory space, and user needs to pay extra money for data storage. In future work we can try to resolve this issue by storing required data based on the user demand and save storage space.

References

1. Pachpor N, Prasad PS (2018) Improving the performance of system in cloud by using selective deduplication. In: IEEE 2018 Second international conference on electronics, communication and aerospace technology (ICECA), 314–318
2. Meyer DT, Bolosky WJ (2011) A study of practical deduplication. In: Proceedings of 9th USENIX conference on file storage technologies, 1–14
3. Koller R, Rangaswami R (2010) I/O deduplication: utilizing content similarity to improve I/O performance. In: Proceedings of USENIX file storage technologies, 1–14
4. Mao B, Jiang H, Suzhen W, Tian L (2015) Leveraging data deduplication to improve the performance of primary storage systems in the cloud. IEEE Trans Comput 65(6):1775–1788
5. Xia W, Jiang H, Feng D, Hua Y (2014) Similarity and locality based indexing for high performance data deduplication. IEEE Trans Comput 64(4):1162–1172
6. Mao B, Hong J, Wu S, Fu Y, Tian L (2012) SAR: SSD assisted restore optimization for deduplication-based storage systems in the cloud. In: 2012 IEEE Seventh international conference on networking, Architecture, and Storage, 328–337
7. El-Shimi A, Kalach R, Kumar A, Ottean A, Li J, Sengupta S (2012) Primary data deduplication—large scale study and system design. In: 2012 {USENIX} Annual technical conference ({USENIX}{ATC} 12), 285–296
8. Koller R, Rangaswami R (2010) I/O deduplication: utilizing content similarity to improve I/O performance. ACM Trans Storage (TOS) 6(3):1–26
9. Schindler J, Shete S, Smith KA (2011) Improving throughput for small disk requests with proximal I/O. In: FAST, 133–147
10. Srinivasan K, Bisson T, Goodson GR, Voruganti K (2012) IDedup: latency-aware, inline data deduplication for primary storage. In Fast 12:1–14
11. Hua Y, Liu X (2012) Scheduling heterogeneous flows with delay-aware deduplication for avionics applications. IEEE Trans Parallel Distrib Syst 23(9):1790–1802
12. Meyer DT, Bolosky WJ (2012) A study of practical deduplication. ACM Trans Storage (ToS) 7(4):1–20
13. Fu M, Feng D, Hua Y, He X, Chen Z, Xia W, Huang F, Liu Q (2014) Accelerating restore and garbage collection in deduplication-based backup systems via exploiting historical information. In: 2014 {USENIX} Annual technical conference, 181–192

14. Wu S, Jiang H, Mao B (2012) {IDO}: Intelligent data outsourcing with improved {RAID} reconstruction performance in large-scale data centers. In: 26th Large installation system administration conference ({LISA} 12), 17–32
15. Zhang X, Wang Y, Wang Q, Zhao X (2019) A new approach to double i/o performance for ceph distributed file system in cloud computing, In: 2019 2nd International conference on data intelligence and security (ICDIS), 68–75
16. Tang Y, Zou Q, Chen J, Li K, Kamhoua CA, Kwiat K, Njilla L (2018) ChainFS: blockchain-secured cloud storage. 11th International conference on cloud computing (CLOUD), 987–990
17. Dang LM, Piran MJ, Han D, Min K, Moon H (2019) A survey on internet of things and cloud computing for healthcare. Electronics 8(7):768
18. Zhang J, Xie N, Zhang X, Yue K, Li W, Kumar D (2018) Machine learning based resource allocation of cloud computing in auction. Comput Mater Continua 56(1):123–135
19. Liao J, Ishikawa Y (2012) Partial replication of metadata to achieve high metadata availability in parallel file systems. In: the Proceedings of 41st international conference on parallel processing (ICPP '12), 168–177
20. Yuan X, Li A (2023) A dynamic chunking algorithm approach for data deduplication, 8th International conference on information systems engineering (ICISE), Dalian, China, 429–432
21. Zhang Q, Song H, Zhou K, Wei J, Xiao C (2024) A prefetching indexing scheme for in-memory database systems. Futur Gener Comput Syst 156:179–190
22. Leconte M, Paschos G, Gkatzikis L, Draief M, Vassilaras S, Chouvardas S (2016) Placing dynamic content in caches with small population. In: IEEE INFOCOM 2016-The 35th annual IEEE international conference on computer communications, 1–9
23. Bhavya M, Pushpa CN, Thriveni J, Venugopal KR (2020) EFUMS: efficient file upload and mutli-keyword search over encrypted cloud data, 11th International conference on computing, communication and networking technologies (ICCCNT), Kharagpur, India, 1–6
24. Sun R, Fan Y, Liu J (2024) A repair-efficient piggybacking design with small number of substripes. IEEE/ACM Trans Networking 32(2):1573–1583
25. Chen R, Huang C, Qin X, Ma N, Pan M, Shen X (2023) Energy efficient and differentially private federated learning via a piggyback approach. IEEE Trans Mob Comput 23:2698–2711
26. Buddesab T, Thriveni J, Venugopal KR (2019) Trust model genetic node recovery based on cloud theory for underwater acoustic sensor network. Int J Electr Comput Eng 1(9):3759

SPRINGER NATURE

GPSR Compliance

The European Union's (EU) General Product Safety Regulation (GPSR) is a set of rules that requires consumer products to be safe and our obligations to ensure this.

If you have any concerns about our products, you can contact us on ProductSafety@springernature.com

In case Publisher is established outside the EU, the EU authorized representative is:

Springer Nature Customer Service Center GmbH
Europaplatz 3
69115 Heidelberg, Germany

The manufacturer's authorised representative in the EU is Springer Nature Customer Service Centre GmbH, Europaplatz 3, 69115 Heidelberg, Germany. If you have any concerns regarding our products, please contact ProductSafety@springernature.com

Printed and bound by CPI Group (UK) Ltd, Croydon, CR0 4YY
01/10/2026
02225858-0005